Preeti Jain
Shankar Lal Garg

Environmental Nanotechnology

Preeti Jain
Shankar Lal Garg

Environmental Nanotechnology

LAP LAMBERT Academic Publishing

Impressum / Imprint

Bibliografische Information der Deutschen Nationalbibliothek: Die Deutsche Nationalbibliothek verzeichnet diese Publikation in der Deutschen Nationalbibliografie; detaillierte bibliografische Daten sind im Internet über http://dnb.d-nb.de abrufbar.

Alle in diesem Buch genannten Marken und Produktnamen unterliegen warenzeichen-, marken- oder patentrechtlichem Schutz bzw. sind Warenzeichen oder eingetragene Warenzeichen der jeweiligen Inhaber. Die Wiedergabe von Marken, Produktnamen, Gebrauchsnamen, Handelsnamen, Warenbezeichnungen u.s.w. in diesem Werk berechtigt auch ohne besondere Kennzeichnung nicht zu der Annahme, dass solche Namen im Sinne der Warenzeichen- und Markenschutzgesetzgebung als frei zu betrachten wären und daher von jedermann benutzt werden dürften.

Bibliographic information published by the Deutsche Nationalbibliothek: The Deutsche Nationalbibliothek lists this publication in the Deutsche Nationalbibliografie; detailed bibliographic data are available in the Internet at http://dnb.d-nb.de.

Any brand names and product names mentioned in this book are subject to trademark, brand or patent protection and are trademarks or registered trademarks of their respective holders. The use of brand names, product names, common names, trade names, product descriptions etc. even without a particular marking in this work is in no way to be construed to mean that such names may be regarded as unrestricted in respect of trademark and brand protection legislation and could thus be used by anyone.

Coverbild / Cover image: www.ingimage.com

Verlag / Publisher:
LAP LAMBERT Academic Publishing
ist ein Imprint der / is a trademark of
OmniScriptum GmbH & Co. KG
Heinrich-Böcking-Str. 6-8, 66121 Saarbrücken, Deutschland / Germany
Email: info@lap-publishing.com

Herstellung: siehe letzte Seite /
Printed at: see last page
ISBN: 978-3-659-74322-1

Copyright © 2015 OmniScriptum GmbH & Co. KG
Alle Rechte vorbehalten. / All rights reserved. Saarbrücken 2015

Contents

Chapter I

1.1	Introduction to Nanotechnology	3
1.2	Nanotechnology Roadmap	8
1.3	Nano-Scale Research and Development Activities: Policy, Ethics and Communication	11
1.4	Relationship between Universities, Research Centers, Industry and Government for Nanotechnology	14
1.5	Role of Mass Media: Journalist and Broadcasters	16

Chapter II

2.1	Kind of Policy Responses to Nanotechnology at National, European and International Level	19
2.2	Social and Political Interests, Values and Institutions Affected By, and Shaping Nano-Scale Developments	23
2.3	Introduction to Ethics	24
2.4	Professional Ethics and Codes	30
2.5	Ethical Issues Involved in Nanomaterials Research	30

Chapter III

3.1	Marketing in Nanotechnology	32
3.2	Understanding, Surveying and Commercializing the Market Place	34

Chapter IV

4.1	Intellectual Property Rights	36
4.2	Intellectual Property Issues Associated with Nanotechnology	37
4.3	Indian Patent Filing Procedure	41
4.4	World Patent Filing (WIPO) Procedure	45

Chapter V

5.1	Environmental and Social Impact of Nanotechnology	49
5.2	Nanotechnology – Health and Safety Issues	54
5.3	Human Exposure of Nanoparticles	55

5.4	Nanoparticles in Aquatic and Terrestrial Environments	57
5.5	Nano-particles in Atmosphere	59
5.6	Health Impacts of Nanoparticles	63
5.7	Toxicological Properties of Nanoparticles	64
5.8	Nanotube Toxicology	67
5.9	Development of Safe Nanotechnology	68
References		72

Chapter I

1.1 Introduction to Nanotechnology
Nanotechnology

Nanotechnology has been emerging as one of the most significant fields all around besides physics, biotechnology. It is a toolbox that provides nanomaterial sized building blocks for the tailoring of new materials, devices & systems.

From chalk to abalone shell, this is the alchemy of natured nanotechnology without human intervention. And now physicists, chemists, materials scientists, biologists, mechanical and electrical engineers & many other specialists are pooling their collective knowledge and tools so that they too can tailor the world on atomic and molecular scales.

Nanotechnology is currently in a very infantile stage. Nanotechnology centers are popping up around the world as more funding is provided & nanotechnology market share is increasing. Nanotechnology could prove to be a transformative technology comparable in its impact to the steam engine in the 18th century; electricity in the 20th century & the internet in contemporary society. The worldwide workforce necessary to support the field of nanotechnology are estimated to be 20 million by 2015. Nanotechnology holds the promise of new solutions to problems that hinder the development of poor countries, especially in relation to health and sanitation, food security & the environment.

Man has learned a lot from nature. Yet his manufacturing practices are primitive. Everyone knows that a lot more needs to be done to get closer to the nature. For, example no one has reached the efficiency of photosynthesis in storing energy. No one can facilitate energy transfer (or electron transfer) as efficiently as biomolecules. No factory does water purification and storage as efficiently as coconut trees or watermelons. The brain of one person can, in principle store and process more information than today's computer. It is unlikely for any movie camera to capture visuals more vividly than the human eye. The olfactory receptors of the dog are much more sensitive than the sensors we have developed through single molecule detectors have been reported. Since the origin of life, communication in one form or the other was there in the history. Even the minutest form of life, such as bacterium & fungi are communicating, though the mechanism is different.

But the technology, especially nanotechnology makes the revolutions in the history. The change was so rapid that no one realized when the pigeon has become electrons. It is really thought provoking that how nanotechnology brings out revolutions in telecommunication, computing, networking and medical industries. It is certain that nanotechnology is penetrating into every aspect of our life & will make the world different from what we know. As sure as the sun rises every day, information and the technology behind it continue to flow and transform at breakneck speed.

Nanotechnology has many definitions since it can be applied in a wide area[1,2,3]. According to US National Nanotechnology Initiative (NNI), nanotechnology consists of three main aspects:

(i) At atomic or molecular scale, technology and research development works.
(ii) Due to their small structures at nanoscale with unique size dependent properties it is used in a variety of applications at that scale.
(iii) At the atomic scale, manipulation and control of materials can be done [4].

The wide extent of scientific studies in this field includes fabrications, understanding and applications of materials, devices and systems at nanometer dimensions. Scientific research about the nature and possible applications of these nano-materials are revolutionizing and can bring much benefit all over the world. Forthcoming developments in nanotechnology, through which impossible can be made possible are nanomaterial with novel optical, electrical & magnetic properties

Today, nanotechnology has encompassed many disciplines such as electronics, computing, medicine, catalysis etc[5].Thus nanotechnology, with attractive properties opened the door to the development of high performance materials. The properties of particles in nano range are strongly dependent on their size, morphology and preparative methods.

The main feature of nanomaterials is their small size in some dimensions; nanomaterials have come to be classified by the number of dimensions in which they are confined to the nanoscale. According to Lojkowski and Fecht classification in nanomaterial is[6]:

(i) Three -dimensional nanostructure (3-D), thick-film-multilayer
(ii) Two-dimensional nanostructure (2-D) coatings, thin-film-multilayer
(iii) One-dimensional nanostructure (1-D) Nanowires, nanorods, nanotubes
(iv) Zero-dimensional nanostructure (0-D) nanoparticles, quantum dots.

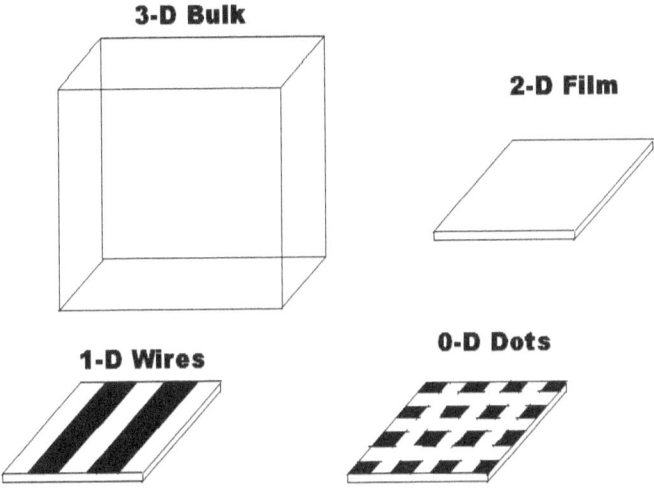

Fig 1.1 Schematic diagram of showing 3-D, 2-D, 1-D, 0-D nanomaterial

Nanoparticles

The introduction of new materials every decade has paved the way for the advancement of knowledge in science. Nanomaterials appear to have taken this development in science to new heights & these materials are expected to revolutionize the application of science in every sector of human endeavour. Nanoparticles have been a steady source of research and development for the last forty years or more. Long before this time, nanoparticles have found uses in such capacities as art. Even today, art galleries and event host displays of paramagnetic ferrofluids being used in free flowing sculpture and design, moving under the influence of electromagnetic control. In recent history, however, the purpose of the nanoparticle has shifted to a more industrialized, medical, or research driven field.

One of the most important branches of nanotechnology is the preparation and usage of nanoparticles. They are small object of nanometer size considered as a whole unit, in terms of its properties. They can be further classified according to size; ultrafine particles are sized between 1 and 100 milimeters. Similarly, as ultrafine particles, nanoparticle ranges between 1 to 100 nanometers.

On the other hand, fine particles cover a range between 100 and 2500 nanometers. Nanoparticles may exhibit size-related properties that differ significantly from those observed in bulk materials (Fig 1.1). Thus, they have drawn great interest at the atomic level as they represent a bridge between bulk materials and molecules[7]. Currently nanoparticles attracted great interest in a variety of applications such as electronics, optical and biological fields.

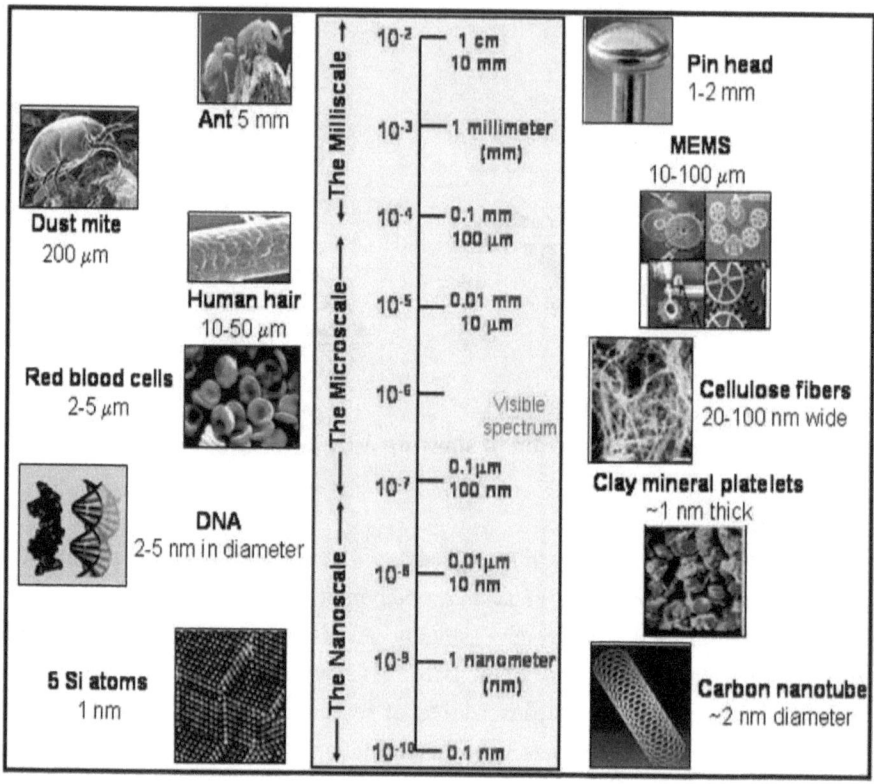

Fig.1.2 A picture representing the relative sizes of various natural and man-made objects

Properties of Nanoparticles

The physical properties of nanostructure materials differ fundamentally from that of the corresponding bulk materials. This is due to the reduced dimensionality on the one hand, and to the modification of fundamental properties on the other hand, as the system size approach quantum mechanical scale, optimization of geometry structure, morphology and the electronic, mechanical and optical properties of nanomaterial sized systems are of fundamental importance for the design of nanostructure with favorable properties. Essentially, the reduction in the particle size of bulk to nanosize results in an increase in the proportion of surface energy & also alters the interparticle spacing[8].

Advantages of Nanoparticles

The current advancement in the field of medicine is the research and development of nanoparticles and nanotechnology for medical application. As a result a new branch of science has emerged. Nanomaterial which is a multidisciplinary field includes biology, chemistry, physics, engineering and materials science. There are several reports and ongoing research on the use of nanotechnology for early detection of several diseases. It has been proved that nanoparticles show a very high sensitivity towards the detection of target bio molecules even when they are present at a very low concentration. Nanoparticles control and sustain release of the drug during the transportation as to achieve the efficiency of drug therapy and reduce side effects. Without any kind of chemical reaction, drugs can be easily injected into the systems without any chemical reaction and thus can be preserved for longer time[9]. Nanotechnology has the potential to bring dramatic improvement in the field of bio-medical sciences including detection, diagnosis & therapeutic systems.

Limitations of Nanoparticles

In spite of above mentioned advantages, nanoparticles do have some limitations.

- Due to their sizes and greater surface area leads to aggregation, that can make physical handling of nanoparticles difficult in dry or liquid forms.
- Nanoparticles because of its large surface area and small particle size readily result in limited drug targeting and its release.
- Before making nanoparticles commercially available in following field viz., nano particulate drug delivery systems, surface modification issues, drug loading strategies, release control and potential applications of nanoparticles. These practical problems have to be overcome first.

Nanotechnology has a huge number and variety of applications across many different sectors. Potentially it could lend to more efficient and have a beneficial impact for the vast majority of people throughout the world. However, as with all technologies, there are also potential negative impacts on society. The main issue includes privacy, social divide, communication and risk.

Risks with nanotechnology

1) Economic disruption from an abundance of cheap products.
2) Economic oppression from artificially inflated prices.
3) Personal risk of criminal or terrorists use.

4) Personal or social risk of abusive restrictions.
5) Social disruption from new products/lifestyles.
6) Unstable arms race.
7) Collective environmental damage from unregulated products.
8) Free-range self replicators (gray goo)
9) A Black market in nanotech (increases other risks)
10) Competing nanotech programs (increases other risks)
11) Attempted relinquishment (increases other risks)

Some of these risks arise from too little regulation & others from too much regulation. Several different kinds of regulation will be necessary in several different fields. An extreme or knee-jerk response to any of these risks will create fertile ground for other risks. The temptation to impose apparently obvious and simple solutions to problems in solution must be avoided.

Nanomaterial will have broad energy implications, but considerable challenges exist regarding the integration of basic research and commercialization. The challenges brought by advanced nanotechnology will have to be addressed by a diverse collection of people and organization. No single approach will solve all problems or address by needs. The only answer is a collective answer & that will demand an unprecedented collaboration, a network of leaders in science, technology, business, government & NGO's.

1.2 Nanotechnology Roadmap

A technology roadmap provides proper information and assessment of the concerned technology. It involves identifying capable technological developments & gaps. It also helps in identifying ways to leverage R&D investments, thus making the technology efficiently to face the competition. It is a good marketing tool as it defines technological evolution in advance.

The process of technology roadmap contains three phase viz., introductory stage, technology roadmap development, and post-development activity[10].

- Introductory stage includes: (1) identify and satisfy all the necessary conditions (identifying the key customers and key suppliers, etc.) (2) provide resource allocation through sponsorship/leadership and setting financial mechanism. (3) The scope of technology roadmapping via the basic requirements, planning and fine details should be defined and also the boundaries should be noted.
- Development of the technology roadmap includes: (1) the "product" that is to be focus, should be identified. (2) Identify the basic needs like the overall framework

and the targets that are to be achieved (3) Explore and identify the major technology. (4) Specify the variables that drive the technology (with targets). (5) Identify alternatives that are available for the technology and the time it will take to mature (6) propose the alternatives of the technology that should be followed. (7) Finally a report is created.

- Follow-up activity includes: (1) the roadmap should be validated by timely monitoring and tracking. (2) Develop an execution plan. (3) Evaluate and refine

Nanotechnology roadmap focuses fast growing developments in nanotechnology landscape and the initiatives that will be undertaken in near future. However, all the views and perspectives of active players should be taken into account before developing such roadmap. Players include different stakeholders-researchers from the scientific and social sciences community, industry, policy makers, development professionals, risk related professionals and civil society etc.

The roadmap for nanotechnology development is a planned strategy which sets a pathway for a product/technology from lab to market. Mostly, the focus is on end-user applications. Direct intervention in augmenting existing sectoral competency, development of nanotechnology based clusters or creating nanotechnology clusters within existing sectoral clusters is examined in defining the new roadmap. Some of the nanotechnology roadmaps in different sectors are given below:

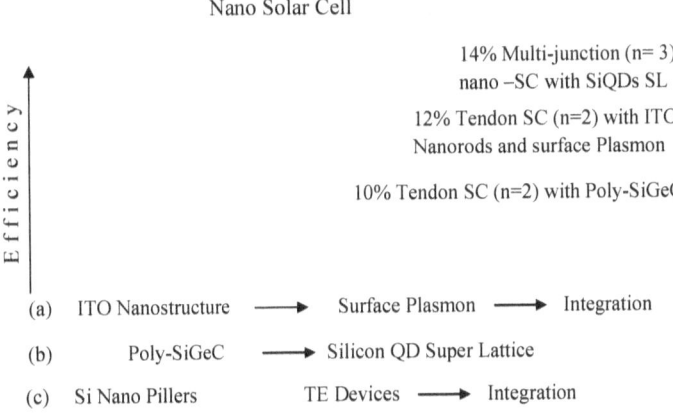

Fig 1.3: Research Roadmap for highly efficient nano-silicon solar cells[11]

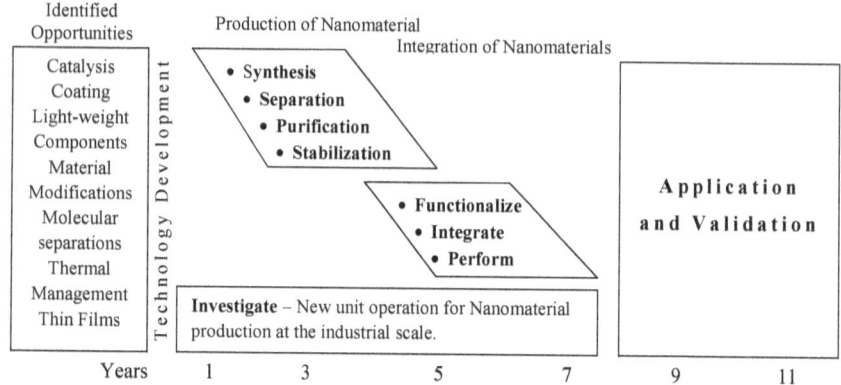

Fig 1.4: Roadmap for Research, Development & Demonstration (RD&D) towards nanotechnology giving large energy savings[12]

(A) Capability of Multi-Function Material

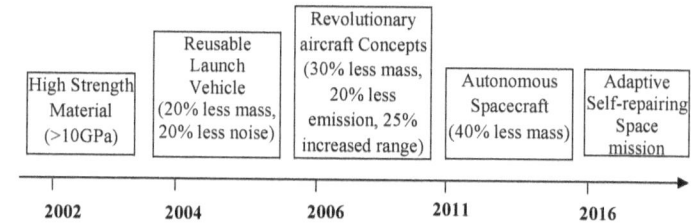

(B) Increasing levels of system design and integration

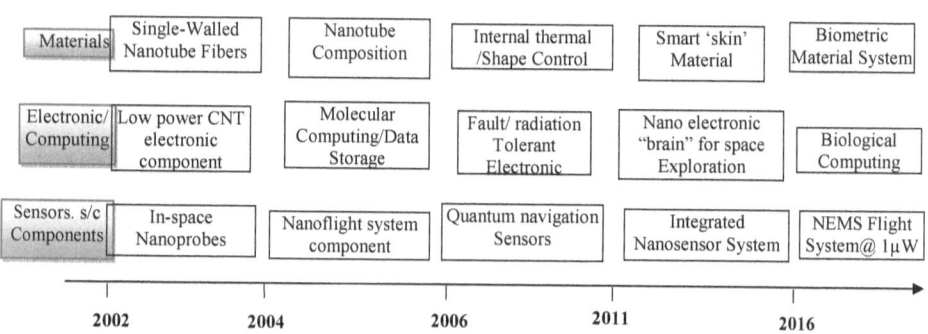

Fig1.5: NASA's Nanotechnology Roadmap[13]

The roadmap demonstrates the RD&D pathway for industrial applications of:
- Catalysis – reactions at lower temperature and byproduct reduction
- Coatings – reduced-drag, self- lubricating and low-friction surfaces
- Light-weighting – low sliding, conveying, and rotating weights
- Material modification – extremely-hard, resistance to wear and improved properties
- Separations – evaporation and distillation process alternatives
- Heat management – superior thermal transfer fluids, reduced conductivity barriers
- Thin films – energy storage, thermo-electric recovery of heat

1.3 Nano-Scale Research and Development Activities: Policy, Ethics and Communication

Nanotechnology seems to be one of the high priority areas of funding due to the 'promises' this technology has demonstrated. Every country has an eye over latest developments happening in this technology. The tag "multi-disciplinary" compels to explore social, ethical, environmental, economic and legal dimensions in addition to various science and engineering sub-disciplines.

On one hand, the field seems to be a blessing by bringing in a revolution in the industry with its mighty wealth and good health while on the other hand, there is increased concern with the possible risks the field might bring in and also the hidden current and future challenges it will throw on us.

For a responsible development to happen, the ethical, legal, and societal implications of nanotechnology must be studied. Pondering over the queries like the means to introduce nanotechnology applications and research into society. The transparency criteria of related decisions taken the sensitivity and openness of the policies designed with respect to the perceptions and needs of stakeholders? The means to tackle ethical, social, and legal issues, are necessary as they will determine the future of the technology and the trust that the public will lay over.

The Policy Framework should intend to improve the quality of life and society, as well as environmental consumption and conservation in a sustainable manner. There are many inherent environmental and medical benefits, besides its manufacturing advantages which help in providing a comfortable lifestyle. For eg., nanomachines, help in better designing and synthesizing pharmaceuticals; to treat cancerous cells and the like, to monitor a patient's health; or to make microscopic operations over the inaccessible parts of the body. As far environment is concerned, nanomachines can clean up oil spills or toxins, eliminate landfills, and recycle all garbage. Having such a versatile application, its potential dangers to eliminate human race on earth cannot be neglected. To say, if, a "Gray Goo Scenario" arose due to some technical fault with nano-disassembled as it starts dismantling every molecule, it comes in contact with and what if the limiting mechanism of self replicating nanomachines get messed up, it would result in nanomachines getting multiplied endlessly like viruses. All those miniaturized microphones, homing beacons and cameras to track and monitor people can rob us from the freedom and privacy. Therefore, the policy framework should work towards nullifying or minimizing the damaging effects on society. This can be done by providing solutions to all the ethical problems before any irreparable harm is caused to the society.

Henceforth, safe and responsible development of nanosciences and nanotechnologies should be adopted. Some of the ethical aspects of technology that crosses with functioning of government are the following:[14]

- Decisions involving which ends should receive priority in funding and allocation of resources in pursuit of those ends whose justification will depend on relative value judgments.
- Regulations are essential be it any technology. They are intended to attain smooth development and accounts for any associated expenses. Regulation has power, oversight, control and responsibility aspects and usually involves allocating risks and benefits. All of these deals with ethical issues and decisions.
- The major responsibility lies with the government who can raise awareness and support research on issues related to ethics and society associated with nanotechnology.

A Typology of ethical and social issues associated with emerging nanotechnologies.

1. **Social Context Issues:** They arise due to the association of nanotechnology with society. For example, is uneven ingress to health care/medical technology, limited information on privacy/security protection, frailty in IP protection systems, educational inequalities, uneven exposure to risks related to environment, unfair tariffs, subsidies and trade agreements and improper protection of consumer safety.

2. **Contested Moral Issues:** They arise due to the interaction of nanotechnology with immoral practices— i.e., those activities which are considered prohibited by a significant number of citizens. For example, those moral practices in which

nanoscience and technology are involved include the development of biological weapons, human enhancements (artificial organism construction), synthetic biology, research on human embryonic stem cell & its associated therapeutics and human beings genetic modifications and human prospecting.

3. **Technoculture Issues:** They arise from the problematic dimension of the increasingly interwoven culture and nanotechnology. For e.g., over-dependency on technology to fix or manage problematic effects (instead approaching the problem at the root), techno-determinism, or the influence of technology over the social changes or the inevitable effects it will bring about, techno-hubris, or a greater belief over technology that it can protect humans from anything etc.

4. **Form of Life Issues:** It arises due to the nanotechnology's potential to have a synergistic effect on social norms, institutions and practices. For example., if nanomedicine lengthens human's average life, even in five healthful years, reconsideration of human development norms will be needed and a significant impacts may be created on family standards (e.g., care taking), trajectories of life (e.g., when to get married) and political and social institutions (example, Medicare).

5. **Transformational Issues:** It is related to the ability of nanotechnology with other technologies like Information technology, biotechnology, robotics etc to transform the dimensions of the human life. It is done by significantly changing the human nature, restoring the relationship humans' had with natural environment or creating artificial persons. In such cases, some important aspects of ethics are to be redesigned—for e.g., the sense of being human, personal identity, etc.

Public understanding of impacts of nanotechnology is necessary. Awareness can be spread among the stakeholders and the public by promoting constructive dialogues and understanding of the potential impact of nanotechnologies. The scientific community and the government agencies that fund scientific research must communicate more directly with the public, rather than using surrogate means like the entertainment industry to demonstrate more clearly to the public the value of nanotechnology.

Large-scale public engagement activities and open meetings focusing on nanotechnology issues should be held that can provide the public with several opportunities to provide input. As we move into the next decade of large-scale funding and the first forays of regulation, it is time to have public engaged in nanotechnology for the benefit public's benefit, and not just for the benefit of nanotechnology.

1.4 Relationship between Universities, Research Centers, Industry and Government for Nanotechnology

A fast growing technology like nanotechnology should maintain tie-up among industry, research centers, universities and government. So as to keep track between the research developments and market outputs. Converting theories and knowledge into product processing is the prime driving force of a country's economic development. A closely weaved relationship between educational institutions and industries will reduce the bridge between the demand of industries and the skill of academician. This will also direct the innovations to grow into job-creating commercial products and services.

Public-private partnerships such as, government's support for technology development and strategic research and enhancing commercialization and technology diffusion is also an important factor to strengthen the economic development.

Great economist like John Hagedoorn, Albert Link, and Coburn has derived the taxonomy of public-private partnership[15]. According to Hagedoorn, based on Members' Composition, he categorized it as (i) public, (ii) private and (iii) public-private and based on Organizational structure, he classified it as (ii) formal comprising research corporation and joint venture research and (ii) informal. Albert Link classified it on the basis of Government involvement as on (i) direct and (ii) indirect while Coburn taxonomies based on Services and Benefits to the industry as (i) Development to the technology (ii) Industrial problem solving (iii) Technology financing (Iv) Start-up assistance and (v) Teaming.

Nanotechnology has potential application in various industrial sectors and so the industry, academia and government collaboration it requires an extended management. To comply, a new approach is established by the U.S. government. A new board (Joint Consultative Board of Nanotechnology) is formed to implement the approach, whose objectives are:

- Collaborative research and planning in areas such as R&D should be encouraged
- R&D results should be publicly displayed encouraging technology transfer
- Research gaps are to be Identified and promoted new in currently running research programmes.
- Information exchange should be encouraged and opportunities should be given industries to propose Research & Development topics

The collaboration may be in the form of co-publication among various educational and research centre as observed with German research institutions in nano science & technology, 1999-2003[16]. More than 75% of the publications of various German universities were in collaboration with Max-Planck Society and Fraunhofer Society, while very little collaboration was observed between the different research centers. Such highly differentiated collaboration can be rationalized in the terms of:

- Research capacity expansion: leading to the combination of expertise and complementary knowledge; access to expensive instrumentation and equipment; and the flair to build project consortia that can compete for third party funding.
- Current research intelligence improvement: keep on improving current research activities by learning new techniques or skills.
- Exchange of complementary among universities and extra-universities in terms of latest knowledge bearing junior researchers and students and access to research topics, facilities and instruments.
- Providing job opportunities to Docs and Post-doc students.

Co-patenting in Nanotechnology

As much as the co-publication network is seen, the co-patenting activity is less observed. Companies and extra research institutes dominate separately in the patent landscape of nanotechnology, but very few occurrences of their collaboration (co-patenting) is observed. A study on co-patenting activities between research institutes and companies in Germany was tried by Heinze[17]. The following are his observations:

- High-profile companies, interact with research institutes that perform fundamental research and applied research. He proposes an intense collaboration between high-technology companies and use-inspired basic research institutes in the development of new nanotechnologies. On comparing the interaction methodology among fundamental research institutes and companies, profound relationships were seen between use-inspired research institutes and high-profile companies in contrast to high-profile companies and basic research institutes.
- Companies typically cooperate with public research units, maintaining a stronger external connection while they are weakly integrated internally. The company-public research unit collaboration enhances knowledge base among company workers.
- Collaboration with multiple research institutes were found to be profiting for companies enhancing their technological capabilities whereas companies void of such ties with public research institutes were progressing much slower.

- Around 40% of universities working with nanotechnology were found collaborating with industries[16]. In the year 2007, among the US universities, Massachusetts Institute of Technology, University of California – Berkeley, Stanford University were leading in co-patenting their technology while IBM Corp., Lucent Technology and Dupont Co Inc were taking the lead among the industries in co-patenting (Source: Wang, 2007)
- Thus, partnership with public and private institutes along with the foundation of strong government supported R&D base will enhance the technology research at a greater speed.

1.5 Role of Mass Media: Journalist and Broadcasters

Be it anything, mass media have an upper hand in influencing public. Wherein, scientific developments are concerned, layman completely depends on the media for information. With technologies like nanotechnology, where only a small percentage of the population is directly aware of ongoing research, it is the mass media that shape public attitudes about the new science. *The current Journalism and Mass Communication Quarterly* by *Scheufele* and *Chul-joo Lee*[18] has reported that those dependants on newspapers and the Internet for scientific information have a better understanding of nanotechnology, but the influence of television news is more on the emotional experience of the viewer than on their understanding. The author suggests that the media has the upper hand in influencing the attitude towards nanotechnology or science at a larger scale. Scheufele put forwards that TV news spends its prime time on focusing personal stories rather than broadcasting the scientific explanation.

Mass media reporting reaches a huge number of people, but the topic dwells mostly on the sensational information or those that involves current debates. The news relates to just the central idea of technical applications, and opportunities & risks associated with the new technologies without giving any further detailing. A main function of media is agenda setting. Agenda building is a process of selecting/ emphasizing certain news/ issues over the others. The factors responsible for the inclusion of certain issues on the agenda are broadly classified into three categories[19].

- **Issue Selection:** The selection of issue by the journalist and editors is based upon media ownership, organizational structures, news production norms, and dogma of journalists (Shoemaker & Reese, 1996).
- **News Values:** In 1965, Galtung & Ruge analyzed international newspaper to find out the factors that influence the journalist to place the news at the top of the agenda. They came up with a list viz., negative stories are rated above the positive

ones, geographical proximity, recent stories, timelines, elite people or state, unexpectedness, exclusivity etc.,

- **Factors extrinsic to the media system:** Issues that are thrust onto the media by influential people through paid communications, press releases and personal connections with journalists also make a column in the newspaper.

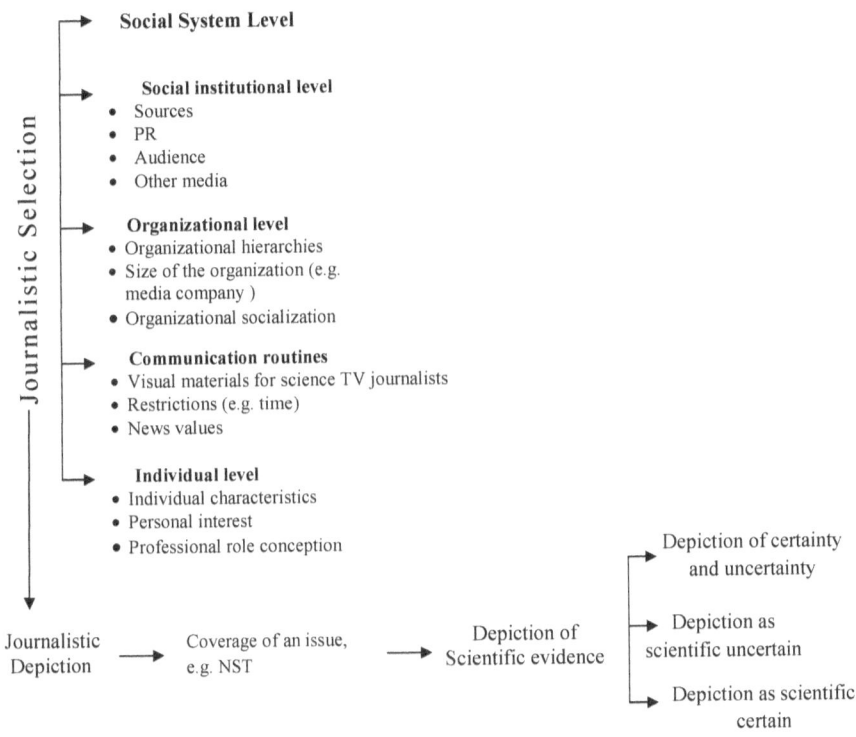

Fig 1.6: Science Journalists' selection criteria and depiction of nanotechnology in German media

Source: Lars Guenther & Georg Ruhrmann, Journal of Science Communication, JCOM 12(03), Autumn/Winter 2013 September 24, 2013
Note: The figure represents two parts: the upper part is focused on the journalistic selection, whereas the lower part focuses on journalistic depiction

Understanding a new technology is difficult for the general public and an extra difficulty is faced by those who do not carry any knowledge of science. And when it comes to communicating risk related messages, the inexpert audience combines all the three factors viz., scientific, emotional and social rather than looking into the real facts. Moreover, personal beliefs and values take the lead role amongst the general public in understanding new technologies and its challenges. Risk related messages have a fine line between overstating and understating which should be considered before communicating. Selection criteria play an important role in science journalism, because it defines the type of news that will be published in the newspaper. Mass media have greatest impact of news issues for naïve audiences as they rely on mass media for information of science to a greater extent. The factors that influence selection in journalism are: the organizational level, individual factors, the social & institutional level, communication practices/routines and the social system level as depicted below[20].

News framing is another key factor for news making and reception. Because the scientific messages communicated should be so framed that public with varied views and values could make out the same actual meaning. Although the journalist prefers to have news frame interpreting significant news, events, the reception studies indicate that the readers go those news events which are emanated from their personal experiences. The following are the criteria which journalist of science newspaper values the most (Anderson et al. 2009, 20–21).

1. Fascination: The news must carry fascinating new items which are not heard before.
2. Size of natural audience: Newspaper readers are more interested events that they are already aware of rather than a rare event.
3. Impact or difference the scientific findings will make in the real world.
4. Timeliness

The Newspaper is among the ones that bring awareness to the general public on emerging technologies, but it happens to journalist conditions. News coverage is basically event-driven practice rather than an issue-driven one, which means the fascination value in real life, takes an upper hand rather than pure scientific findings.

Chapter II

2.1 Kind of Policy Responses to Nanotechnology at National, European and International Level

Department for Science and Technology (DST) of India is striving hard to promote nanotechnology & nano-science for more than 15 years. Under the direction of Professor C. N. Rao (Father of nanotechnology policy in India), Nano Science and Technology Initiative (NSTI), a national funding programme was launched by DST during 2001. DST is involved in numerous nanoscience and nanotechnology activities as follows:

- R & D Projects to Individual Scientists
- Strengthening of Characterization Facilities by establishing sophisticated equipments such as Optical Tweezer, Nano Indentor, Transmission Electron Microscope (TEM), Atomic Force Microscope (AFM), Scanning Tunneling Microscope (STM), Matrix Assisted Laser Desorption Time of Flight Mass Spectrometer (MALDI TOF MS), Microarray Spotter & Scanner etc. at various locations in the country.
- Establishment of Centers of Excellence including eleven Units/Core Groups on Nanoscience housing some of the most sophisticated facilities which are shared with other scientists in the region, seven centers for Nano Technology focusing on the development of specific applications and a center of excellence on Computational Materials Science.
- International Collaborative Programmes with EU, Russia, Ukraine, Japan, Germany and USA.
- Joint Institution-Industry Linked Projects and Public Private Partnership activities.
- Human Resource Development in Nano Science & Technology.

Nanomission, a primary funding body was set up under DST, steered by a Nano Mission Council (NMC), wherein the two advisory groups, viz. The Nano Science Advisory Group (NSAG) and the Nano Applications and Technology Advisory Group (NATAG) takes care of nanomission's technical programmes.

The Nano Mission looks upon the overall development in the nanotechnology and nanoscience and also identifies the applied potentials lies in the country for the nation's benefit. In brief, the objectives of the Nano-Mission are[21]:

- Funding and promoting research (Individual scientist/group)

- Developing the infrastructure necessary for Nano Science & Technology Research.
- Nano Applications and Technology Development Programmes to promote application oriented R&D Projects, establish Nano Applications and Technology Development Centers, Nanotechnology Business Incubators etc.
- Developing or producing Human Resource by initiating M.Sc./M.Tech. programmes, creating national and overseas post-doctoral fellowships.
- International Collaborations in terms of joint research projects, workshops, conferences and sophisticated research facilities abroad.

In different Indian state the scenario of policies, research and development in nanotechnology greatly vary. Some states are progressively contributing to the domain while others are idle. Karnataka is among the leading states followed by Andhra Pradesh in nanotech research. Bengaluru, Karnataka's capital city seeks to become "Nano City of India". Similarly, the state of Andhra Pradesh also hosts ICICI Knowledge Park (IKP) in Hyderabad housing various research programmes in nanotechnologies.

There seems to be very less funding for Nanomaterial Risk assessment in India. There are only four institutes currently work over nanotoxicology, which promotes safety related Nanomaterial biomedical research in India viz., the Indian Institute of Toxicology Research (IITR), the National Institute of Pharmaceutical Education and Research (NIPER), the Indian Institute of Chemical Technology (IICT), and the Central Drug Research Institute (CDRI). In addition, the Indian Council of Medical Research (ICMR) formulates. Though in the upcoming years major funding over toxicology studies on titanium dioxide, zinc, silver and carbon may be observed. As few institutes (the Central Food Technology Research Institute (CFTRI), the National Environmental Engineering Research Institute (NEERI), the National Chemical Laboratory (NCL), the National Institute of Oceanography (NIO), the Technology Information, Forecasting and Assessment Council (TIFAC), and the Indian Council of Agricultural Research (ICAR)) who are presently researching over the new nanomaterial's beneficial effects may at one point of time cover its toxicological effects.[22]

Risk management: As far as India is concerned, no regulations or laws exist for Nanomaterials. A non-profit research institute (The Energy and Resource Institute (TERI)) is running a project on "Capability, Governance and Nanotechnology Developments: A Focus on India" which has multiple aims to promote nanotechnology in a smooth way. One among its aim also holds a placard to make government to have a closer look over regulatory issues in nanotechnology. So far, DST in 2011 has appointed a task force, which reports the Nano Mission Council, to work on the terms of reference and other details of a regulatory body for nanotechnology.

Key areas of the European Strategy and the Action Plan

Research: Interdisciplinary approach for research in nanotechnology is promoted through collaborative R & D across Europe by bringing public and private institutes together. Crucial research topics like safety of nanoparticles, pre-normative research, or research for health, security, energy, information society and environment. The less developed countries and socially disadvantaged people are identified and supported.

Industrial Innovation: In small scale and medium scale enterprises, it is important to have a favorable environment for innovations to happen. Nanotechnology has to pay special attention to factors involving research and development, regulation, metrology and patenting the knowledge, skilled human resources, functioning and competitive markets, public-private partnerships, financial instruments and infrastructure.

Infrastructures: European Union aims at providing the best available infrastructure comprising both single sited and distributed facilities. Because of the complex interdisciplinary nature of nanotechnology the infrastructure requires a huge funding, which is beyond the means of industry and national government.

Education: The very basis of human resources. Europe aims at providing high quality up-to-date education at the same time interdisciplinary training and promoting collaborative research programmes among academia and industry. They have made attempts to attract youths, especially women and girls towards science and nanotechnology in particular.

Societal aspects / Ethics: European commission has made a forum for open debate wherein interested people can share their views. Using which risk, benefits and ethical issues (both real and perceived) are analyzed that will help in directing the technology in a responsible manner.

Risk assessment: The risk assessment has given special attention. European commission encourages research towards potential impacts of nanotechnology on health and the environment through promoting toxicological and ecotoxicological studies.

Regulation: Regulatory frameworks are examined for their applicability to nanomaterials in order to propose adaptations of EU regulations in relevant sectors.

International cooperation / International dialogue[23]
Policy responses to nanotechnology at International level

United States of America: In 2001, USA launched National Nanotechnology Initiatives to promote R & D and USA's stand in Nanotechnology. In 2003, it enacted '21st Century Nanotechnology Research and Development Act' which established programs, assigns responsibilities to various agencies, authorizes funding level for agencies and imitates research works.

NNI offers funds on the processes required for nano manufacturing, advances in understanding basic nanoscale phenomena, developing nanoscale systems and devices, nanomaterials, standards, instrumentation, measurement science and the tools. NNI also fund for basic infrastructure and facilities required for nanotech R&D. Finally, it supports risk assessment research, along with its legal, ethical, and societal implications.

Structure: The NNI is coordinated within the White House through the National Science and Technology Council's NSET subcommittee. The NSET subcommittee is comprised of representatives from federal agencies, White House Office of Science and Technology Policy (OSTP), and Office of Management and Budget. The NSET subcommittee has established four working groups: the National Environmental and Health Implications (NEHI), National Innovation and Liaison with Industry (NILI), Global Issues in Nanotechnology (GIN), Nanomanufacturing and Nanotechnology Public Engagement and Communications (NPEC) working groups. The National Nanotechnology Coordination Office (NNCO) provides administrative and technical support to the NSET subcommittee[24]

Many other countries and industries are also trying their hands in investing nanotechnology to obtain its potential benefits. Around 60 countries have manifested nanotechnology programs between the year 2001 and 2004[25].

China: The National High Technology Research and Development Program, known as the 863 program looks after China's science and technology including nanotech and science. It aims at promoting new materials and manufacturing technologies in various fields including Nanomaterials. Its first project "Climbing Project on Nanomaterial Science" was a great success, recognizing which the government restructured its funding program towards nanoscience. A combined funding of USD 30 Million has been received by 10 nanotechnology projects since

2006, under the program overseen by the Ministry of Science and Technology (MOST). Under Medium and long term development plan (MLP), nanotechnology is considered as Science's mega project and is given high priority status.

In the year 2000, the National Steering Committee for Nanoscience and Nanotechnology (NSCNN) was established to look after and co-ordinate the various projects in nanotech and science. The NSCNN is directed by Dr. Chunli Bai and it consists of MOST, the Chinese Academy of Sciences (CAS), National Natural Science Foundation (NSFC), the National Development and Reform Commission (NDRC), the Ministry of Education (MOE) and the Chinese Academy of Engineering (CAE) (interviews, NSCNN, March and October 2010)[26].

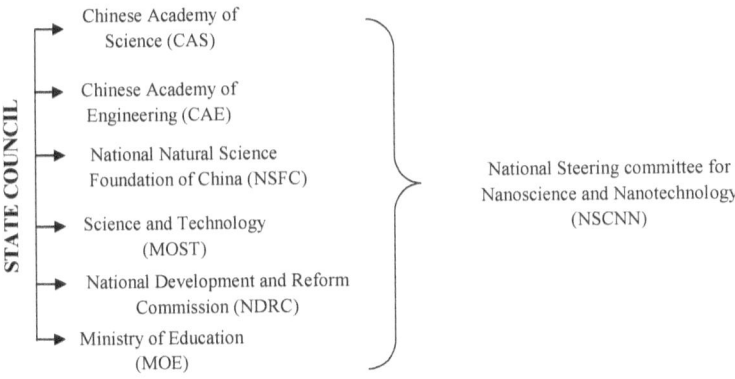

Fig 2.1: Science and Technology planning in China

2.2 Social and Political Interests, Values and Institutions Affected By, and Shaping Nano-Scale Developments

The shape or the future of any new technology in coming years is also governed by social and various non-technical aspects. Social and political interest often contradict each other. Nanotechnology's development in security and surveillance/ privacy field can affect the country's security and its political stability, if the technology goes into the hand of offensive groups like terrorists. This will disturb the sense of balance between its offensive and defensive abilities, Any kind of privacy breaches may directly affect social values.

Suppose a pervasive monitoring and surveillance systems is developed solely for military purposes, there are possibility that the use of such persistent surveillance technology may not be restricted to military related activities. The technology may further get decentralized to state government who for security reason may gather data of foreign individuals or government or may gather intelligence information on particular sections of the population. This could result in political protests, or the public may experience interference in their privacy[27].

2.3 Introduction to Ethics

The word 'Ethics' is coined from the Latin word 'Ethicus' and the Greek word 'Ethikos' means science of conduct, manner and character. This meaning can also be extended to imply systematizing, defending and recommending concept of virtues, right and wrong behaviour. Ethics is thus said to be the science of moral, morality, moral principle and recognized rules of conduct and human behaviour. Ethics is the branch of philosophy which is the systematic study of selective choice of the standards of right and wrong and by which it may ultimately be directed for actions.

or

Ethics is a system of moral principle governing the conduct and behaviour of individuals and groups.

or

Ethics is the outcome of combined study of religion, social science and civic sense which governs the conduct and behaviour of individuals and groups based on their value judgment, belief, moral and perceptions.

or

Ethics is a branch of philosophy. Its object is the study of both moral and immoral behaviour in order to make a well founded judgment to arrive at adequate recommendations. For example moral acts include honesty, truthfulness, kindness to animals, etc. Immoral act mean theft, murder, telling a lie etc.

Nature of Ethics

1) Ethics is a subject that deals with human beings. Human by their nature are capable of judging between right and wrong,
2) Ethics deal with human conduct that is voluntary and not forced by any person or circumstances.
3) Ethics is more a science and not an art because it is a systematic knowledge about moral behaviour and conduct of human beings.
4) Ethics affirms an unchanged moral code and needs the indignation of the spirit to give added impact to its force on behalf of the right privileges and dignity of man.

5) The outcome of ethical activities is judged based on the long range strategic planning of business.

Objectives of Ethical study:

The study of ethics is nothing but a field of philosophy and social science in which a set of systematic knowledge about value judgment, moral behaviour and human conduct is learnt and tested. It also assesses whether a particular action or decision taken by an individual is moral or immoral. The benefit of the study of ethics includes the following:

- Bring clarity in thought and help in reasoning and decision making.
- Reduce and avoid stress and strain in day to day working.
- Establish moral standards and norms of behaviour to build character.
- Set code of conduct for professionals.
- Adjust to adhere to organizational value.
- Apply judgment measures upon human behaviour based on standard and norms.
- Suggest moral behaviour and prescribe recommendations about what is right or wrong, good or Evil, true or false, fair or unfair, proper or improper, ethical or unethical and about Do's and Don'ts.

Ethical Situation:

In any profession or business or even in our day to day activities, we come across many diverse and conflicting situations, where decision makers are totally confused in deciding what is right or what is wrong? What is ethical or what is unethical? Based on varieties of internal and external moral pressures.

Example 1:

The CMD of a paper mill may dislike cutting of trees (To protect the environment) but as an executive his feelings should not interfere with the best interest of the company, since without tree cutting paper cannot be produced. What are your recommendations?

Example 2:

The son of a high court judge has committed a murder. The judge as a parent may feel an obligation to protect his son from criminal proceedings. But as a law abiding citizen, on the other hand, the same judge may feel obligated to co-operate with the police and court in prosecuting his child. What do you suggest under such situation?

Example 3:

When sales of finished products of a manufacturing company dropped, the supervisor is asked to terminate a few employees to save cost. The supervisor knows that some of them though loyal and hard working have to be terminated to save his skin. To him the ethical situation relates to two dimensions:

1. Personal 2. Professional

How will you handle such situation?

Ethical Situations for a professional:

- A customer asked for a product from the shopkeeper. After telling him the price, he said he could not afford it. I know he could get it cheaper from a competitor. Should I tell him about the competitor or let him go without getting what he needs? What is the guideline for shopkeeper? The societal interest or the personal interest?
- My computer operator told me that he had noticed several personal letters printed from a computer that I was responsible to manage. While we had no specific policies against personal use of company facilities, I was concerned. I approached the letter writer to discuss the situation. She told me she had written the letters on her own time to practice using our word processor. What should I do?
- A fellow employee told me that he plans to quit the company in two months and start a new job which has been guaranteed to him. Meanwhile, my boss told me that he was not going to give me a new opportunity in our company because he was going to give it to my fellow employee now. What should I do?

Morals and Morality

The Moral is the principle of right or wrong. Moral values are deep seated belief, feeling and norms that manifest themselves as behaviour or conduct of a person. These values are neither measurable nor expressed in words.

or

Morals are a set of rules of conduct and standards of evolution that a culture uses to guide its individual and collective behaviour and direct its judgment.

Morality is derived from the Latin word 'moralis' which also refers to the customs of a social group. Morality is the code of value to guide the actions and behaviour of an individual human being. Morality is a primary force in shaping ethics and the development of ethics is dependent on the religious morality.

Moral standards: Moral standards deal with desired right or wrong human behaviour. They focus on what ought to be done and what ought not to be done by an individual or group in a given situation and time.

Morality vs Ethics

Really speaking to differentiate between ethics and morality is a difficult task, as human behaviour is influenced by emotions, sentiments, perceptions and beliefs.

However Ethics and Morality seem to be synonymous, but it is not so exacting. Morality is a primary force in shaping ethics. Morality is concerned with the standard of conduct and motives, right and wrong or good and bad character. Ethics is the philosophical study of morality. The development of Ethics is also dependent on the religious morality. Ethics is not merely the code of conduct based on customs, conventions and the accepted courtesies of a society, but it is the code of conduct developed by proper testing to guide the human behaviour.

Various Ethical Views, Criteria and Codes:

Philosophers and ethicists after lifetimes of thoughts are unable to agree on a simple definition of ethics. However, their concepts fall into the following categories which an individual can use for ethical reasoning:

1) Utilitarian based Ethics
2) Right based ethics
3) Duty based ethics
4) Virtue based ethics

Utilitarian based Ethics:

This ethical concept was developed by 'Jerry Bentham' and 'John Stuart Mill' of England. Utilitarian based ethics say that an action is right from an ethical point of view if and only if the sum total of utilities produced by that act is greater than the sum total of utilities produced by any other act which could have been performed in its place.

or

Utilitarian based ethics states that an action is normally correct if it produces the greatest benefit for the greatest number of persons which is guided by the value "Utility". Decisions are purely made on the basis of their outcomes or consequences. The duration, intensity and equality of distribution of benefit should also be considered. Decision makers choose utilitarian criteria while making the decision for termination, closing down plants, laying off a large number of employees in the best interest of the organization. Some utilitarian advocate the use of cost/ benefit analysis in evaluating specific courses of action. They maintain that the course of action that produces the greatest benefit relative to cost is the one that should be chosen. Benefit is usually defined in some relatively specific way, such as producing jobs or something else of value to society.

Right based Ethics:

All human beings on earth are free and equal. Right based ethics are based on the belief that there are certain fundamental human rights like health, liberty, possessions and the product of his or her labour and those moral obligations arise in the context of these rights.

For example, 'Whistle blowing' is the latest phenomenon occurring in the modern corporate world of the 21st century. If decision makers use right criteria, a good protection can be given to whistle blowers when they blow against some wrongdoers.

Duty based Ethics:

It states that every person has a duty to follow those courses of action that would be acceptable as universal principles for everyone to follow.

Virtue based Ethics:

This concept emphasizes the role of individual personality traits. It explores these traits and qualities that would help the individual lead a better life. Virtuous acts are done willingly and not by chance. Both ethical principles and virtues need to be considered together.

or

It is concerned with attaining dispositions of personality that an individual desires in himself or others.

or

Virtue based Ethics states that inner happiness is to be achieved by developing virtues or qualities of character, through deductions and reasoning. An act is good if it is in accordance with reason.

or

Virtuous people who lack ethical principles are ethically blind, but ethical principles without virtuous those people are empty.

Unethical Human Behaviour and Practices in Business and Society are as follows:
1) Deterioration of human belief, values, customs and character
2) Corruption and Fraud
3) Favoritism
4) Nepotism
5) Loss of faith in laws, courts and government
6) Violation of local laws and regulation by multinational corporation (MNC)
7) Employments of children in general utility and services
8) Non compliance with environmental ethics and issues
9) Financial irregularities
10) Tax evasion
11) Business crimes

12) Gambling on asset values
13) Stock holding and adulteration
14) Black marketing
15) Poor quality of products and services
16) Kickbacks (bribes offered by vendors to buyers)
17) Underworld business of drugs and crimes

Business Ethics

Business ethics are concerned with moral issues in business and commercial transactions that define right and wrong behaviour in the world of business, just as medical ethics are concerned with morality of medical practices and policies, or political ethics are concerned with morality of political affairs.

Business ethics has various aspects viz.,

- Fair competition and affordable cost of goods
- Customer satisfaction, joy and pleasure
- Corporate behaviour
- Protection of consumer's right
- Excellent payments of wages and salary to employees
- Social responsibility and better relationships.

Professional Ethics

Professional ethics are the applications of general ethical rules, code of conduct and behavioural pattern to various professions like Medical, Engineering, Pharmacy, IT, Computer, Management, Finance and Charted Accountancy, Fashion designing etc.

Professional Responsibility:

Professional responsibility encompasses specific virtues like:

1) Self direction and self introspection
2) Proficiency, integrity, self respect, honesty, truthfulness and impartial decision
3) Honesty in beliefs, speech and acts
4) Sensitivity to confidentiality
5) Commitment to society and team spirit to the enhancement of human welfare and safety
6) Supporting the professional and technical societies of their disciplines
7) Avoiding unsafe ventures

Engineering Profession:

Engineering is a highly privileged profession. It elevates the standard of living, luxuries and adds comfort to the human life.

Engineering Vs Other Professions

"An engineer cannot bury his mistakes in the grave like the doctors. He cannot argue them like Lawyers and he cannot screen his shortcomings by blaming his opponents and others."

Engineering Ethics:

Engineering Ethics is defined as "The study of the moral issues, character, policies and decisions confronting individuals and organizations engaged in engineering activities. Engineering ethics are the widely accepted codes and standards of conduct by engineering organizations and societies.

2.4 Professional Ethics and Codes

Professional Codes:

A professional code is set of rules, regulations and standards which are accepted as general principles which state how professionals in a particular organization or society should behave. Codes of ethics basically reflect an organizational primary value, norms, beliefs and ethical rules of operations.

List of few important professional societies and their codes of ethics for engineers

1) American Society of mechanical engineers (ASME)
2) Institute of electrical and electronic engineers (IEEE)
3) National Society of professional Engineers (NSPE)
4) Accrediting board of Engineering Technology (ABET)
5) Indian Society of Technical Education (ISTE)
6) Computer Society of India (CSI)
7) Institute of Engineers (India)
8) Association of Computing Machinery (ACM)

With every new developing technology like nanotechnology, there are unknown ethical issues that can arise. It is purely the researchers' responsibility to look for two preliminary issues. The first is does the ethical issues raised by nanotechnology are unique and, if not, should the researchers still worry about the ethical issues it raises.

2.5 Ethical Issues Involved in Nanomaterials Research

The second is will examine a list of ethical responsibilities will be enough to have a check over the ethical issues it may raise. Scrutinizing research ethics is a very crucial part of research in Nanomaterials. It is also the responsibility of researchers to discuss and find out the ethical implication their research work may have on the society. NT researchers whose work is publicly funded have an ethical responsibility to society at large to do the best work they can to generate reliable new scientific and engineering knowledge, materials, devices, and systems[28].

Chapter III

3.1 Marketing in Nanotechnology

Nanotechnology covers a broad area of expertise and so it cannot be defined as "an industry" like "an automobile industry". Owing to the above statement, *Business Week* quotes "Nano is not a single industry, but a scale of engineering involving matter between 1 and 100 nanometers". Nanotechnology had very few applications at the incipient stage. Now, however, the disparate engineering and science disciplines' convergence has paved way to numerous applications in various spheres such as computer chips, biotechnology, materials, manufacturing, security, energy, medical diagnosis and health care, space exploration and so on. Thus, nanotechnology will have an effect on every such firm that it is dealing with. So, such a vast technology is unlikely to succeed without appropriate research into understanding and surveying the market place.

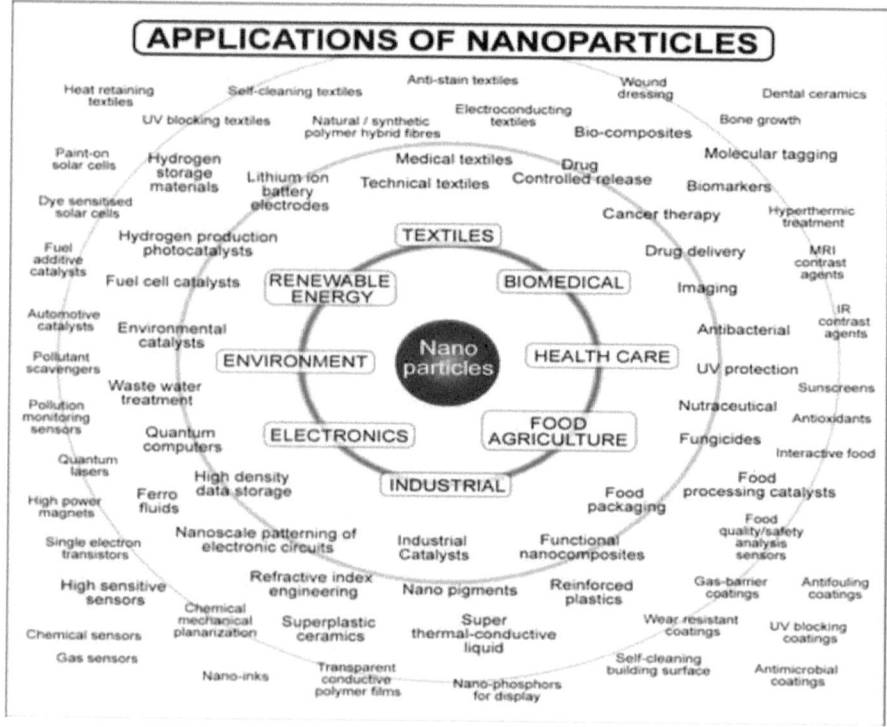

Fig 3.1: Application of Nanoparticles

Source: Tsuzuki T., 2009, Commercial scale production of inorganic nanoparticles. International Journal of Nanotechnology, 6, 567-578

The market for emerging nanotechnologies has grown exponentially over the past decade. As per the BCC research report on nanotechnology, by 2013, the nanotechnology's global market was valued at $22.9 billion and by 2014 it reached $26 billion[29]. The total sales are expected to reach $64.2billion in 2019 with an increase of compound annual growth rate (CAGR) of 19.8%. Its contribution to global GDP is estimated at $1 trillion by 2020[30]. Allied Market Research[31] predicts the market of global quantum dots to stretch from $316 million in 2013 to $5.04 billion by 2020, at a Compound Annual Growth Rate of 29.9%.

"The New Market Research report on Nanotechnology"[32], states that the majority of the nanotechnology-based products' revenue is shared by the chemical industry. At 58.5% CAGR over the analysis period of 2013-2020, healthcare & pharmaceutical industry seems to be the fastest growing sector. This is because of the high demand in market for nanodiagnostic products, its extensive application of nanomedicines and its waxing usage in cancer treatment methods.

A sizeable investment is required in the seed phase of new product and process developments. The lengthy process of 'concept to commercialization' requires a considerably long term commitment and better financial assistance. The observed significant reduction in prices in nanotechnology based products and materials, will help in broadening its applications over a broad gamut of industries. This will further help in uplifting of the market. Betterment in production units and its number along with increase in healthy collaboration among various institutes, universities and corporate will promote further advancements in the industry.

So much of a progress is found happening in nanotechnology that to understand its market as one industry is not easy. Furthermore, less availability of proper definitions, intricate applications, undefined market boundaries, make it even more unfavorable. Converting ideas into a product demands a lot of skill set in setting up an industry viz., dealing with legal compliances, infrastructure, employer training and hiring, marketing the products etc. Before investing or commercializing any product a fair market survey is required.

Commercialization is the process of introducing new products or technologies into the general market. It takes into account financial, marketing, and technical array to achieve commercial success. This process involves discovery/invention of a technology/product, evaluation of the invention, its market evaluation, demonstrating the product (developing a prototype), further development & optimization of the technology & product, verification and validation, regulatory approval, introduction of the product/technology into the market, promotion and post-marketing activities[33].

3.2 Understanding, Surveying and Commercializing the Market Place

It is known that Nanotechnology has the capability to create novel, user-friendly industrial and consumer products. Effective tools like market survey and technology roadmap help in better understanding the market. Market survey will give a better picture of the market demand, state-of-art technologies or any intellectual property issues associated with the commercialization. This will provide information relating to the availability of relevant resources, production facilities and infrastructure in the market that will boost the production of the new technology. In addition, it will shed light on concealed environmental issues, health and safety hazards that are emerging from the processing and utilization of nanomaterial, in a way making commercialization a convenient stuff.

The survey also helps to identify the risk factor and uncertainties involved in the products' commercialization. It gives an idea about the business readiness of the product and its timely launch.

The survey further explores governments' support and resource availability to form collaboration in pursuance of offering a safe and profitable commercial use of the technology. When a complete background survey of the product related area is done the next step is to commercialize the product.

The two key factors involving commercialization of a product is to have a strong IP portfolio while establishing a company and secondly a clear, risk-free and cost-effective business plan.

1. Protecting IP is done by filing of patent or licensing the patent from universities or by government labs.
2. Gathering funds which may be through government funding or private funding.

Other factors that involve in commercialization of a product are:

- Strengthening R & D by giving freedom to perform research of their (researchers) interest and to follow interesting leads. Funding the research project should be based on criteria that will result in high-impact and successful research. The criteria, adopted by the European Union are based on three main aspects: (i) The quality of Science and technology. (ii) Implementation of the research findings (iii) Impact of the research. Having a proper balance between the three types of research viz., applied, platform and fundamental, will enable a better research environment.
- Using an improved infrastructure.

- Associating Research with Industrial and Economic Strategy: The major industries that have a hand in the nanotechnology research are the following:
  a. Construction
  b. Defense & Security Systems
  c. Health
  d. Mining and Materials
  e. Oil & Gas
  f. Petrochemical
  g. Telecom Infrastructure
  h. Transportation, Water

Specific sub-fields within the principle industries that involve nanotechnology researches are:
a. Corrosion resistance (Petrochemicals, Oil & Gas, Defense, Construction & Infrastructure, Water, Mining & Materials).
b. Deep drilling developments (Oil & Gas).
c. photonic and electronic nanodevices, NEMS/MEMS (Oil, Defense, Industry, Water, Transportation and Medical)
d. Enhanced catalysis (Petrochemicals)
e. Enhanced oil recovery (Oil & Gas)
f. Enhanced well productivity (Oil & Gas)
g. Improved desalination (Water)
h. Medical diagnosis & drug delivery
i. Monitoring nanodevices (Defense, Water, and Medical)
j. Renewable energy such as solar cells (Industry, Water)

- The International Collaboration Plan will promote streamlined nanotechnology research and will alleviate the standard of knowledge among the researchers by decreasing the knowledge gap between developed and developing countries.
- Sound business plan by identifying potential market for each product via market analysis. Availability of vibrant resources (e.g., human resources, expert advisor, financial experts, etc.) will further enhance the process of commercialization.
- Monitoring environment and health issues[34]

Chapter IV

4.1 Intellectual Property Rights

They are the exclusive rights granted to the creator(s) and owner(s) of a work such as inventions; literary and artistic works; designs; and symbols, names and images used in commerce that are the creation of mind. Intellectual property rights include Patents, Copyrights and related rights, Trade Marks, Geographical Indications, Industrial Designs, Layout Designs of Integrated Circuits, Protection of Undisclosed Information (Trade Secrets), Maskworks and Plant varieties.

A Patent is an exclusive right granted by a country to the inventor to exclude others from making, using, offering for sale, or selling the invention throughout the country or importing it into the country for a limited time (say 15 to 25 years) in exchange for public disclosure of the invention when the patent is granted. A Trademark ™ is a word, phrase, symbol or design, or a combination thereof, that identifies and distinguishes the source of the goods of one party from those of others. A Copyright © protects works of authorship, such as writings, music, and works of art that have been tangibly expressed. A Service Mark "SM" is a word, phrase, symbol or design, or a combination thereof, that identifies and distinguishes the source of a service rather than goods. The term "trademark" is often used to refer to both trademarks and service marks. Geographical Indications are indications that identify Goods referring to a country or to a place situated therein as being the country or place of origin of that product like Mysore silk, Tirupathi Laddu, Puneri Pagadi, Bikaneri Bhujia, and Chanderi Fabric etc. An Industrial Design constitutes the ornamental or aesthetic aspect of an article such as watches, jewelry, architectural structures, etc. A design may consist of three-dimensional features, such as the shape or surface of an article, or of two-dimensional features, such as patterns, lines or color. A Trade Secret is any information that allows you to make money because it is not generally known. A trade secret could be a formula, computer program, process, method, device, technique, pricing information, customer lists or other non-public information. If the economic value of a piece of information relies on it being kept private, it could be a trade secret. For example, the formula for Coca-Cola, which is referred by the code name "Merchandise 7X". Maskwork (Layout Design Of Integrated Circuit): In chip technology, when the chip layout includes an original circuit design, the layout is protectable. Specifically, the maskwork protect against the unauthorized copying of the chip layout information. A Plant Variety Rights are a grant that gives the breeder the exclusive right to produce for sale and sell propagating material of the variety. In the case of vegetatively-propagated fruit, ornamental and vegetable varieties, Plant Variety Rights give you the additional exclusive commercial right to propagate the protected variety for the commercial production of fruit, flowers or other products of the variety[35].

4.2 Intellectual Property Issues Associated with Nanotechnology

Nanotechnology involves research at the atomic or molecular scale (1-100nm) of materials, devices and systems that exhibit physical, chemical and biological properties that are different from those found in larger scales. Thus, nanotechnology can be better understood as a broad collection of technologies -- from diverse fields such as physics, materials science, engineering, chemistry, biochemistry, medicine and optics - each of which may have different characteristics and applications.

Patent Applicability: As it is known, the properties of matter and other fundamental scientific discoveries are not patentable and for a patent to be granted the idea/invention should be novel, useful and non-obvious. Hence, simply submitting a smaller version of a known structure would not be considered, as the sole element of novelty is a difference in size, since a mere change in size may be viewed as 'obvious'. In fact, patents have been refused even in situations where the change in form, proportion, or size brought about better results than the previous invention, which is common for nanotechnology. Hence, by drafting patent applications, presenting the invention as a solution to these new problems at the nanoscale rather than giving importance to the size, will give inventors a better chance of obtaining a patent.

Balancing Innovative Freedom and Restrictive Intellectual Property: The increasing rate of patent applications from universities and private research organizations highlights another potential challenge for the nanotechnology industry. The most basic ideas in nanotechnology are already patented or may well end up being patented by universities at large. These building block patents can be very lucrative because the fundamental technologies they claim may become prerequisites for many downstream innovations. Before commercializing nanotechnology products, companies may have to obtain licenses from a large number of patent owners. In order to attain the proper balance between innovation and exclusion, patent strategists will need to consider ethical questions about the division and aggregation of legal rights and reassess the scope of licensing practices.

The possibility of overlapping claims to intellectual property is always a risk. Here, the risk is particularly critical because many of the patents that have been and are currently being filed are building-block patents—fundamental patents that can affect the development of future innovations and products to market. Where parties obtain overlapping protections, a patent thicket can result: no one can develop a product without infringing another's patent, and no one want to "give up" rights to allow another to develop a product.

Patent office Challenges: Due to the multidisciplinary nature of the subject, its inventions are based on a wide spectrum of technologies, including materials science, electronics, physics, chemistry and biology.

While this diversity may foster hurdles to the examiners, as they may not be familiar with advances in other areas necessary for the complete examination of a new technology, creating significant difficulties in patent examination, classification, and analysis.

For example, the broad definition of 'nanotechnology' leads to challenges in classifying new inventions for Patent Office purposes. On one hand, an application may use other terms, such as 'microscale' or 'quantum dot,' to describe a nanotechnology invention. On the other hand, an applicant may incorrectly describe his invention as 'nanotechnology,' perhaps seeking to capitalize on the positive press associated with this term, or use terms like 'nano-second' that arise in other contexts. Inventors and examiners must, therefore, be particularly cautious when searching for prior art in this area—'nano' alone is not a good search term. Because nanotechnology encompasses such a broad swath of disciplines, often teams of dedicated, specialized professionals must work in concert to further research and development. The scope and breadth of nanotechnology provide opportunities to develop numerous cross-use applications and products. Complex functions can be achieved in a host of unrelated applications.

As a result of these definitional difficulties, securing intellectual property protection is difficult. One consequence of such a multidisciplinary emerging field has been the awarding of overlapping interests of different entities, thereby creating conflicting claims of right. Where, for example, patent examiners are not sure which "art" category to search, or where they employ limited keyword search terminology, inconsistent review of patent applications is inevitable. Where entities do not know what makes "Nanotech" different, infringement of pre-existing rights is a distinct possibility.

Foreign Patents: Patent protection is typically only effective within the issuing country. In light of the considerable worldwide efforts in nanotechnology research, early foreign patent protection will be essential. Securing international patents will increase the administrative effort and expense of nanotechnology patent protection. This can be achieved by filing a PCT application which includes a specific nanotechnology classification (IPC Class B82B). The PCT ability to defer nationalization and formal examination for 30 months from filing also offers companies two and a half years to further develop and examine the value and viability of the inventions disclosed in applications to determine whether the inventions merit the further costs of nationalized prosecution. Whatever the technology, it is critical to those who decide to pursue patent protection that a well-developed plan is generated to consider and address the difficulties associated with nanotechnology filings both domestically and abroad.

Trade Secret Challenges: Trade secret protection offers the advantage of avoiding the effort and expense of patent applications. It has a potentially indefinite duration, subject, of course, to reverse engineering. With lengthy commercialization timelines for some nanotechnologies. The 20-year limit of the patent term, it may be advisable to opt for trade secret protection as long as the product is not easy to reverse engineer in the near future. However, trade secret protection requires continuous diligence, and once a trade secret is revealed, it has no further protective value. Pressure to publish in academic circles makes the trade secrets difficult to maintain. It is very difficult to obtain government funding and maintain trade secrets given the governmental funding reporting requirements. The increase in funding and companies pursuing nanotechnology applications further will increase employee mobility and necessitate stringent safeguards against the theft of trade secrets by departing employees. Finally, as a general matter, investors tend to avoid technologies that lack patent protection, making trade secret protection a non-viable option for many innovative technology companies.

Intellectual Property Litigation: Nanotechnology intellectual property litigation has already emerged around trade secret issues. For example, in July 2000, Caliper Technologies Corporation ("Caliper") sued Aclara BioSciences ("Alcara") for misappropriation and conversion of Caliper's proprietary technical, strategic and intellectual property information relating to microfluidics. In response, Aclara sued Caliper for patent infringement. After Caliper obtained a jury verdict against Aclara in its trade secret suit the parties settled. Later in October 2002, Nanogen announced the settlement of a lawsuit with former employee Donald Montgomery for taking its trade secrets to Acacia Research Corporation's ("Acacia") CombiMatrix unit and filing patent applications related to the disputed technology under his name. Under terms of the settlement, Acacia agreed to pay Nanogen $1 million to cover litigation costs and issue 4 million shares, or 17.5 percent, of its unit's stock. Acacia also will pay Nanogen royalty payments on sales of products developed by either CombiMatrix or affiliates that use the disputed technology. Finally, Zyvex Corporation, a company developing NanoElectroMechanical Systems ("NEMS") for prototype nanoscale assemblers, obtained a permanent injunction against a former employee for misappropriation of trade secrets.

Trademarks: Companies in the Nanotech space must be careful to avoid descriptive branding, as descriptive terms are difficult to trademark. For example, names such as "nanosilver" and "nanoparticle" tend to be descriptive and are difficult to protect. Developing a fanciful mark that becomes synonymous with innovative nanotech designs and products is far more defensible. Moreover, to the extent that regulations concerning the presence of nanoparticles in consumer goods begin to change, the prefix "nano" may be included in a certification seal required under other regulations. Such developments can affect the value of a trademark. Trademark counsel who is aware of the developments outside the Patent and Trademark Office, including of the Environmental Protection Agency and the Food and Drug Administration, can assist companies in securing rights to defensible marks. Once trademark protection is secured, companies must be vigilant in policing the unauthorized use of their marks to maximize the value and protect the goodwill associated with them. Especially in an area where the presence of nanotechnological applications and devices cannot be readily discerned by the public, companies that invest in the technology and market products should protect their marks from being hijacked by competitors or imposters.

Conclusion: Nanotechnology is an emerging technology with exciting prospects for intellectual property, in both the near term and for many years to come. The importance of securing and maintaining intellectual property should be recognized by all the players. In light of the massive influx of funding from the Nanotechnology Act, strategists must consider the impact of government sponsorship in conjunction with the ever present time and cost constraints that will shape strategies for protecting nanotechnology intellectual property[36].

4.3 Indian Patent Filing Procedure

The Patent System in India is governed by the Patents Act, 1970 (No. 39 of 1970) as amended by the Patents (Amendment) Act, 2005 and the Patents Rules, 2003, as amended by the Patents (Amendment) Rules 2006 effective from 05-05-2006.

Who can apply? The inventor may make an application, either alone or jointly with another, or his/their assignee or legal representative of any deceased inventor or his assignee. If an application is filed by the assignee, proof of assignment has to be submitted along with the application. The applicant can be national of any country.

How to file?: A patent application shall be filed online or at an appropriate office along with Provisional / Complete Specification, with the prescribed fee.

A Patent application is of four types. They are a) Ordinary Application b) Application for Patent of Addition (granted for Improvement or Modification of the already patented invention, for an unexpired term of the main patent). c) Divisional Application (in case of plurality of the inventions disclosed in the main application). d) Convention application, e) National Phase Application under PCT.

Convention application: India is a member of Paris convention, WTO agreement, Budapest Treaty and Patent Co-operation Treaty (PCT) – any country which is a member of the convention/treaty/agreement are convention countries. When an application is filed in any Convention country (basic application), the applicant can make an application for a patent within twelve months after the date on which the basic application was made, claiming the priority date on the basis of filing in Convention Countries.

PCT application: An Indian applicant can file a PCT International application by filing directly with the International Bureau of WIPO after taking permission from the Indian Patent Office (IPO) or filing anytime from 6 weeks to 12 months from the date of filing at IB of WIPO. However, international filing within six weeks requires permission from IPO. In some cases there is probability that permission may be deferred and the application may be referred to DRDO / Department of Atomic Energy for their directions. On the other hand an international patent application can be filed directly at Indian Patent Office to request for permission to do so at IPO.

Usually a complete specification should including (i) title, (ii) abstract, (iii) description, (iv) drawing (if any), (v) sample or model (if required by the examiner), (vi) enablement and Best Mode, (vii) claims and (viii) deposits (microorganism) is submitted while a provisional application, including (i) title, (ii) description, (iii) drawing (if any) and (iv) Sample or model (if required by the examiner), can be filed at a stage where some experimentation is required to perfect the invention. However, a provisional specification cannot be filed in case of a Convention Application (either directly or through PCT routes). A complete specification should be filed within a year's time from the date of filing the provisional application.

Priority date: It is the first filing date of an application. In case of provisional filing, a provisional application is followed by a complete application; the priority date will be the filing date of provisional application and not the complete one. It is similar to the PCT application too. The importance of priority date is that it sometimes becomes the prime factor in granting patent as the first filed applicant receives the patent grant.

The following documents should be submitted at the time of filing (i) Application for the grant of patent (ii) Provisional or Complete application, (iii) Statement and undertaking by the applicant, (iv) Declaration as to inventorship and (v) Authorization of a patent agent or any other person. Priority document details have to be filed for a Convention application.

The Territorial jurisdiction of a patent office: Territorial jurisdiction of a patent office is decided based on the following: i) Place of residence, domicile or business of the applicant (first mentioned applicant in the case of joint applicants). ii) Place from where the invention actually originated. iii) Address for service in India given by the applicant, in case of foreign applicants.

Territorial jurisdictions are presented below:

Patent Office	Territorial Jurisdiction
Mumbai	The States of Gujarat, Maharashtra, Madhya Pradesh, Goa, Chhattisgarh, the Union Territories of Daman & Diu and Dadra & Nagar Haveli.
Delhi	The States of Haryana, Himachal Pradesh, Jammu and Kashmir, Punjab, Rajasthan, Uttar Pradesh, Uttarakhand, National Capital Territory of Delhi and the Union Territory of Chandigarh.
Chennai	The States of Andhra Pradesh, Karnataka, Kerala, Tamil Nadu and the Union Territories of Pondicherry and Lakshadweep.
Kolkata	Rest of India (States of Bihar, Orissa, West Bengal, Sikkim, Assam, Meghalaya, Manipur, Tripura, Nagaland, Arunachal Pradesh and Union Territory of Andaman and Nicobar Islands)

In case of filing a divisional patent application, it shall be filed at the same office where the main application is filed, as a divisional application needs to be examined vis-à-vis its main application, while the foreign applicant is required to give an address for service in India and the jurisdiction will be accordingly decided.

Publication: A patent application is published after eighteen months from the priority date. Once published, the rights of the patentee start from the date of publication but they cannot be forced until after patent grant. The application can be published earlier if a request is made by the applicant and will not be published if the direction is given to secrecy, until the directions expires.

Examination: A request for examination has to be made within 36 months from priority or filing date. But in case of secrecy, a request is made six months after the direction is revoked or 36 months from priority or filing date, if that date is later.

On receiving the request, the controller shall direct the patent application to the Examiner for examination. To start with, the examiner makes a formal examination by verifying the propriety and correctness of all documents filed with the application. Later, he verifies the patentability of the application. The patentability analysis includes all patentability requirements. After confirming that the application falls within the scope of patentable subject matter, the examiner conducts a prior art search to check if there is prior art, which anticipates the invention claimed. Prior art search for anticipation includes search for anticipation by publication, filing of the complete specification, etc. He then verifies the existence of inventive step, Industrial application and Enablement and Best mode.

The examiner will give the examination report within one month from the date of reference by controller and that term shall not exceed three months. If the examination report is adverse, the controller sends a notice to the applicant and gives them an opportunity to correct and if necessary, an opportunity of hearing. The Controller might ask the applicant to amend the application in order to proceed further. If the applicant does not make such changes, the application might be rejected. The Controller has the power to divide the application, post date the application, substitute applicants and reject the application. An order of division will be given if the application contains more than one invention and if it is required to file separate applications for each invention. The application might be posted dated to a period of six months if requested by the applicant. Substitution of inventors is generally done if the inventor has been wrongfully mentioned or if a joint inventor has not been mentioned in the application. The controller has the power to reject the application, if the applicant does not comply with his requirements.

Opposition:

1. Pre-grant Opposition: Any person can file an opposition for grant of patent after the application has been published. Opposition may be filed on any of the following grounds.

    (a) Non compliance with patentability requirements

    (b) Nondisclosure or Wrongful disclosure of genetic resources or traditional knowledge

2. Post-grant Opposition: Any person can file an opposition within a period of twelve months after the grant of a patent. It can be filed based on the following grounds: a. Wrongful obtainment of the invention by the inventor. b. Publication of the claimed invention before the priority date. c. Sale or Import of the invention before the priority date. d. Public use or display of the invention. e. The invention doesn't satisfy the patentability requirements. f. Disclosure of false information to patent office. g. Application of the invention is not filed within twelve months from the date of convention application. h. Nondisclosure or wrongful disclosure of the biological source. i. Invention is anticipated by traditional knowledge.

3. Process of Opposition: On receiving a notice of opposition, the controller notifies the patentee. He then constitutes an Opposition board to deal with the opposition. The Opposition board decides the issues after giving reasonable opportunity of hearing to both the parties. The Opposition board might invalidate the patent, require amendments or maintain the status quo. If amendments are required, they have to be made within the prescribed period in order to maintain the patent.

Grant: If the application satisfies all the requirements of the patent act, the application is said to be in order to grant. An application in order for grant shall be granted expeditiously. A granted patent shall be published in the official gazette and shall be open for public inspection. Every granted patent shall be given the filing date. The patent will be valid throughout India. A granted patent gives the patent holder the exclusive right to make, use, sell, offer or sale and import the product or use the process. However, the government can make use of the patent for its own purposes or for distributing an invention relating to medicine to hospitals and dispensaries. Furthermore, any person can make use of the patent for experiment or education[37].

4.4 World Patent Filing (WIPO) Procedure

The PCT (Patent Co-operation Treaty) is an international treaty, administered by the World Intellectual Property Organization (WIPO), between 148 Paris Convention countries. The PCT assist applicants to simultaneously seek patent protection internationally for their invention in about 148 countries throughout the world without the need of filing several separate national or regional patent applications. Though granting of patents remains under the control of the national or regional patent Offices in what is called the "national phase".

An international patent application may be filed by anyone who is a national or resident of a member country. A single patent application can be filed directly with the International Bureau as receiving Office by mail or hand-delivered to WIPO's headquarters or by facsimile provided that the original of the faxed application is furnished within 14 days from the date of the fax transmission, or by electronic filing (PCT Electronic Filing). This single, uniform patent application is what is referred to as the international application.

Filing an international patent application to start the patent process is a wise move if one is contemplating securing patent rights in multiple countries. The PCT filing only removes the burden of filing a patent application in individual countries. Having an international application does not mean that an international patent will be granted. It is not so because there no "international patent" as such. If the applicant has to obtain a patent in a particular country, the international patent application has to be filed in that country. This can be accomplished through the PCT process by entering the so-called "national stage" in the countries where he wants to receive a patent, or the applicant can file a patent application claiming the benefit of international patent application directly in a particular country within 12 months of filing the international patent application.

An applicant can first file a national or regional patent application with their national or regional patent Office, and within 12 months from the filing date of that first application, he can file their international application under the PCT. This international application will hold the priority date as of the first national application filed date. Then the international application can enter into national phase in 30 months from the filing date of the initial patent application of which the applicant claims priority

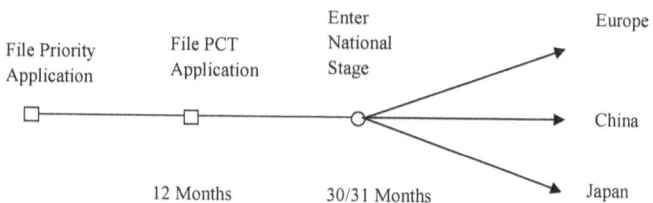

Figure 4.1 Flow Diagram of PCT Application

An international application must be filed in an authorized Receiving Office (e.g, IPO in India) or with the International Bureau as receiving Office in any language. The Receiving Office functions as the filing and formalities review organization for international applications. However, the request has to be filed in one of the ten publication languages under the PCT, that is, in Arabic, Chinese, English, French, German, Japanese, Korean, Portuguese, Russian or Spanish.

International search: The following offices are appointed by the PCT Contracting States as International Searching Authorities (ISAs) the National Offices of Australia, Austria, Brazil, Canada, China, Egypt, Finland, India, Israel, Japan, the Republic of Korea, the Russian Federation, Spain, Sweden and the United States of America, and the following regional Offices, the European Patent Office and the Nordic Patent Institute. If the application is in a language which is not accepted by the International Searching Authority, the applicant need to furnish a translation of the application for the purposes of international search.

An international search report and the written opinion will be received by the applicant from the ISA by the fourth or fifth month after the filing of the international patent application. If the report is favorable, it will assist the applicant in further processing of the application in those countries in which the applicant wishes to obtain protection. But, if a search report is unfavorable, then the applicant can amend the claims in international patent application to better distinguish the invention from the documents that challenge the novelty and/or inventive step of the invention, and have them published, or to withdraw the application before it is published. The high quality of the international search assures you that any patent granted from an international application is less likely to be successfully challenged, and thus provides valuable input in support of investment decisions.

With the search report a written opinion is also sent to the applicant and to WIPO stating whether the invention appears to meet the patentability criteria in light of the search report results, which especially helps the applicant in evaluating the chances of obtaining a patent. A supplementary international search can also be requested by the applicant so as to reduce the risk of new patent documents and other technical literature being discovered in the national phase because it may enlarge the linguistic and technical scope of the documentation searched.

Publication: WIPO publishes the international application shortly after the expiration of 18 months from the priority date (if it has not been withdrawn earlier), together with the international search report. PCT international applications are published online on PATENTSCOPE.

National phase:

To enter the national phase, some requirements should be fulfilled, including paying national fees and, in some cases, filing translations of the application, appointment of local agents. These steps must be taken, in relation to the majority of PCT Contracting States' patent Offices, before the end of the 30th month from the priority date. On entering the national phase, the concerned national or regional patent offices begin the process of determining whether they will grant the applicant a patent.

Advantages of filing a PCT application

The procedure under the PCT has many advantages for the applicant, for the patent Offices and for the general public:

i. The applicant has up to 18 months more than if he had not used the PCT to reflect on the desirability of seeking protection in foreign countries, to appoint local patent agents in each foreign country, to prepare the necessary translations and to pay the national fees.

ii. If the international application is in the form prescribed by the PCT, it cannot be rejected on formal grounds by any PCT Contracting State patent Office during the national phase of the processing of the application.

iii. The international search report and written opinion contain important information about the potential patentability of the invention, providing a strong basis for the applicant to make business decisions about how to proceed.

iv. The applicant has the possibility during the optional international preliminary examination to amend the international application, enter into dialogue with the

v. The search and examination work of patent Offices in the national phase can be considerably reduced or eliminated thanks to the international search report, the written opinion and, where applicable, the international preliminary report on patentability that accompany the international application.

vi. The applicant may be able to fast-track examination procedures in the national phase in Contracting States that have PCT-Patent Prosecution Highway agreements with the Offices which act as ISAs.

vii. Since each international application is published together with an international search report, third parties are in a better position to formulate a well-founded opinion about the potential patentability of the claimed invention.

viii. For the applicant, international publication online puts the world on notice of the application. He may also highlight his interest in concluding licensing agreements on PATENTSCOPE, which can be an effective means of advertising and looking for potential licensees, and

ix. The applicant also achieves other savings in communications, postal and translations because the work done during the international processing is generally not repeated before each Office.

Ultimately, the PCT:
- Brings the world within reach.
- Postpones the major costs associated with international patent protection.
- Provides a strong basis for patenting decisions
- Is used by the world's major corporations, research[38].

(Note: top of page begins) examiner to fully argue the case and put the application in order before processing by the various national patent Offices.

Chapter V

5.1 Environmental and Social Impact of Nanotechnology

Nanotechnology can be precisely defined as the purposeful manipulation of matter at the atomic level and molecular scale. Its versatile properties have found application in various fields such as cosmetics, food, household appliances, cellular phones, computers, medicines, ceramics, textiles, construction materials, military weapons, and sports equipment. Though it is not completely a new technology, recent findings in this field are so advanced that their impact examination upon the world is very much necessary. All these advances carry a potential impact on the environment and society.

Environmental impacts of nanotechnology:

When the environment is concerned nanotechnology itself has come to its rescue, say it was in enhancing the efficiencies of pollution monitoring devices, giving remediation to environmental pollution, or boosting renewable energy production. But that doesn't mean nanotechnology has helped environment in all ways, the increased toxicological pollution on the environment due to the uncertain shape, size, and chemical compositions of some of the nanomaterials is a major concern.

Nanomaterials are introduced into the environment at three stages viz., (i) production (ii) application and (iii) weathering. Having a bigger surface area than the bulk material, the damage caused by them to the environment will also be greater. Naturally occurring nanoparticles have a very little effect on health and environmental. This new technology has a great deal of unknown factors making it more complex to understand its impact at the time of production. The existing technology do not support in detecting the nanoparticles in air as there is less knowledge on its characterization and its nature. The nature of the nanoparticles drastically changes even with a minor change in its chemical structure.

To know the inadvertent consequences of nanotechnology (or any technology) researchers must assess the entire life cycle of the technology, i.e., from the raw materials/process used, operation till its disposal. Otherwise a wrong notion of the extent of pollution will be formed among the user

Table 5.1: Detailed sorting of nanoparticles existing in environment[39]

Environmental Nanoparticles				
Natural Nanoparticles			Engineered Nanoparticles	
Atmospheric	Terrestrial and Aquatic		Unintentional	Intentional
Inorganic (e.g. volcanic ash)	Inorganic • Silicate (e.g. Clay, Mica) • Oxide/hydroxide (e.g. MnO) • Carbonate (e.g. CaCO$_3$) • Phosphate • Metal Sulphite (e.g. ZnS)	Organic • Macromolecule (e.g. HAs) • Bio-collide(e.g. Bacteria) • Cellular debris	• Wearing and corrosion • Wear & combustion Products	• Carbonaceous NPs (e.g. SWCNTs) • Metal Oxide (e.g. ZnO$_2$) • Semiconductor material (e.g. QD) • Zero-valent Metals • Nanopolymer (e.g. dendrimers

Most of the nanomaterials are disposed as incineration and landfill or during wastewater treatment. For example, about 95% of cosmetic and paint nanoparticles end up in the wastewater through abrasion and application during its lifetime. Some of the nanomaterials such as carbon nanotubes, remain intact and leaches into the underground water or released into the air through landfills and incinerators[40].

Once in the environment the nanoparticles may interact differently: they may reach underwater via a carrier such as contaminants, organic compounds or by bio-uptake. These aggregated nanoparticles may get transported to sensitive environment and might break into colloidal nanoparticles. The following are means through which the nanomaterials can become toxic and harmful to the surrounding environment:

i. **Hydrophobic and hydrophilic nanoparticles:** Currently TiO$_2$ nano powder is used as a coating material for composite to reduce weathering effects. As these Ivana Fenoglio, et al. expressed their concern that the effect of TiO$_2$ nanoparticles to be assessed when leaked into the environment.
ii. **Mobility of contaminants:** Nanoparticle are emitted either directly (Primary emission) or indirectly (via nucleation with sulfuric acid and ammonia – Secondary emission) into the atmosphere. Nanoparticles can easily be transported to an environment such as aqueous environments when they get attached to contaminants. They may bio-uptake by organisms or aggregated and deposited

over macromolecules or adsorbed or desorbed by organism. In such a way, it was evident that nanosize nuclear waste traveled almost one mile from a nuclear test site in 30 years.

iii. **Solubility:** Most of the nanoparticles are water soluble and if handled inappropriately they may pollute water which may go irreparable.

iv. **Disposal:** Waste materials, including nanomaterials are major environmental concerns if not disposed safely.

Some common examples of day-to-day uses of nanomaterials

- Graphene has some outstanding properties and finds application in many areas. They are used in replacement of heavy metals/composites reducing the weight of the instruments in which they are used like graphene composites in airplanes, as fire retardants, structural improvement, etc. But if it gets exposed to the environment, it could react to living and non-living systems in an unexpected way as its toxic property are not yet identified.
- Carbon nanotubes are used in drug delivery system, carbon-lithium batteries, as a catalyst in a fuel cell, in sensors. They also find application in pharmaceuticals, textiles, food packaging, etc. When these nanoparticles leaches off into vegetative land they are said to show a low crop yield. Its exposure to rice field has shown a month delayed flowering in rice. Seeds exposed to C70 fullerenes for two weeks may pass it onto the next generation. In wheat plants they pierce through the cell wall make a "path"/"pipe" for the pollutants to flow into the living cells.
- Nano-silver is known for its antibacterial quality and finds application in wound dressings, washing machines and clothes. Few people have shown concern over incidental extension of its application to water treatment plants and to ecosystem as it may kill beneficial microbes.
- Nano ZnO in cosmetics, especially sunscreen renders transparency of the product as its nano-size is transparent to visible light. This nanosized ZnO was found to show toxicity towards vertebrates and bacterias.
- Because of high surface-to-volume ratio, certain nanomaterials are highly reactive. This reactivity promotes them to be used in catalysis, site remediation and treatment of groundwater contamination. However, this reactivity is seemed to have extended to cause cell damage in animals.
- Most of the nanomaterials used in cosmetics, personal care, lubricants, paints/coatings, food packaging, Agrochemicals, health care etc. enter the environment through surface water and sewage. While, few nanomaterials find its way through the atmosphere and soil[40]. For e.g., Gold, nanoclays, the silver used

in food packaging enters through surface water and soil. Cerium oxide, Platinum, Molybdenum trioxide used as catalyst and lubricants sewage enters through sewage and surface water.

Social impacts of nanotechnology:

The changing society is though not direct consequences of scientific discoveries, but they pave the way for the change that arises to fulfill the social and economic needs. Nanotechnology involves varied and extensive application that their effects will take years to work their way through the socio-economic system. Nano-sceptics believe that nanotech will worsen the problems arising from existing inequality distribution among the rich and poor, widening the differences farther through nano-divide; warping the status of international power via its trading and military applications; entrusting ubiquitous surveillance having significant effects on civil liberty; introducing unknown and difficult to understood risks to human health and environment; breaking down the barriers between life and non-life, and redefining even what it means to be human[41].

Social inequality will rise higher as the nano-divide will be developed initially between the nano-poor (world's poorest countries) and the nano-rich countries (world's rich countries that can get along with monetary demands the technology required). Wealthy countries using their available funds will innovate and patent the technology early, thus establishing a control over the technology while the poorer countries may find difficulty in competing with non-nano products with the finished nano-goods. A nano-divide may emerge within the country, as a larger difference between those who have a hold over the new nanotechnologies and those who don't and also affordability criteria will pave way to greater nano-divide. If ever the idea of enhancing humans using nanotechnology is materialized, then a growing difference between the mental, physical and "performance" abilities of people will be observed.

The initial applications of nano-products were in cosmetics or accessory applications such as odor-eating socks, anti-ageing cosmetics; superior display screens for televisions, computers, and mobile phones; self-cleaning windows and bathroom; premium coatings for luxury cars etc. These applications were affordable only by the selected few. Even in the healthcare industry, expensive nano-based treatments were initially launched, providing access only to the wealthy ones. All of this might accelerate negative social impacts

A higher ratio of research in nanotechnology is military funded and finds directed towards military applications which may give rise to lethal high-tech "nano-weapons". In times of war if they are used by both the opposing country, then the result it will produce is unseen, having greater social costs.

"Surveillance", a progress in electronics and information technology in combination with nanotechnology have helped governments to keep track of their citizens. But if the technology goes into the hands of corporations (governmental & non-governmental) ran by heinous people there are chances of abusing the technology by pervasive surveillance of individuals.

A new concern has aroused in nanotechnology – nanobots, a self-replicating nano machines. Although there are a lot of complication and technical challenges that have to be dealt with to make it work in the natural environment. Like, the nanobots if manufactured, that it is an uncertain scenario, should be able to survive the environment they will encounter, should carry a sufficient-sized computer that will store all the data that is required for the bots to work and self-replicate

. But if such an unforeseen task emerges, robots constructing a number of robots and dismantling the old one, similar to the biological system, the changes it might bring about is concealed.

Another danger is molecular manufacturing ("personal nanofactory containing fabricators, manipulators, and an assembler that perfectly builds complex products by assembling an atom by atom at specified place"). If it comes into reality, then a disruptive impact will be observed in labor markets and global trade. A country may develop a lot of destructive weapons quickly at a cheaper rate and easier to hide and surveillance too will become cheaper and easier will pose risks from criminals and terrorist. A larger amount of product at cheap rate will be manufactured that may bring down the economy. There may be significant social disruption because as this personally owned nanofactory can be used for generating banned products. Low-manufacturing price may pose risks to the environment. The need of labor in manufacturing unit will reduce to zero as desktop molecular factories are self-sufficient[42].

Other potential issues like national and international political affairs, including technological leadership, balancing the needs and responsibility, intellectual property protection like what should be patented and what not, media coverage and presentation and queries regarding government, industry and the university's relationship.

An effective and cost-efficient way to protect the public and deal with nanotechnology's potential negative consequences is to develop a tradition of social-science-based countermeasures and to support research in publicly recognized institutions on the processes that develop nanotechnology and apply it in diverse areas of life.

5.2 Nanotechnology – Health and Safety Issues

Nanotechnology is believed to have important societal and economic benefits, but at the cost of potential adverse safety and health issues. Few of the nanotechnology products can induce new or unseen human health risks. For example, nanosize particles may cross bloodstream more readily after ingestion, inhalation, or dermal exposure. The resulting toxicological effects may vary on the basis of the means of exposure to the same material. Likewise, as the size shrinks the explosion and fire hazards associated may increase.

Nanotechnology is regarded as double-edged sword, exhibiting exceptional beneficial properties, as well as frightening unknown consequences, such as environmental and new toxicological effects. Some examples are as follows:

- It is observed that certain nanoparticles can penetrate through the blood-brain barrier, which separates the blood from extracellular brain fluids, thereby protecting the brain from any harmful substances that may be present in the blood and also blocks the therapeutic agents' delivery. The characteristics of these nanomaterials may be used for treating brain damages by purposefully delivering the therapeutics directly to the brain. Though there are chances, that some nanoparticles might cause harm to animals and humans by accidentally penetrating through the blood-brain barrier.
- Silver nanoparticles are said to have effective anti-bacterial properties, but also shows toxicity towards mammalian cell cultures. It is observed in various cells in rats, which are used as models of human cell. There are evidences regarding significant effects of silver nanoparticles on male reproductive system as they show toxicity towards mammalian stem cells. Reports on the detrimental effects on human health due to nanosilver ingestion either directly or through medical devices are observed.
- Carbon nanotubes (CNTs) display a wide range of applications in many fields like materials, memory devices, batteries, electronic displays, sensors, transparent conductors, and medical imaging. Like asbestos fibers, CNTs also harm humans and animals by getting lodged in organs (e.g., kidneys, lungs, livers). Though carbon is less water soluble but its allotrope buckyballs have shown substantial water solubility, and an amount of 500ppb in fish brain has found to cause

peroxidation of lipid after 48 hours. Single-wall nanotube causes free radical formation and depletion of antioxidants in cells.
- Human skin cells when exposed to 0.6-0.24 g/mL unrefined single wall nanotube containing 30% iron, result in oxidative stress, atherosclerosis, Parkinson's disease, Heart Failure, Alzheimer's disease.
- Intratracheal exposure of mice to a single dose of 1, 2.5, or 5mg of multi-wall carbon nanotube results in pulmonary effects, then death[40].

5.3 Human Exposure of Nanoparticles

Some of the health impacts of nanotechnology are

❖ Nanoparticles, which are, incidentally, inhaled goes into the lungs, causing decrement of lung function and fibrosis and obstructive lung diseases and those that enter into the blood stream ends up in other organs. The effects of these studies of engineered nanoparticles, with different particle properties, are uncertain.

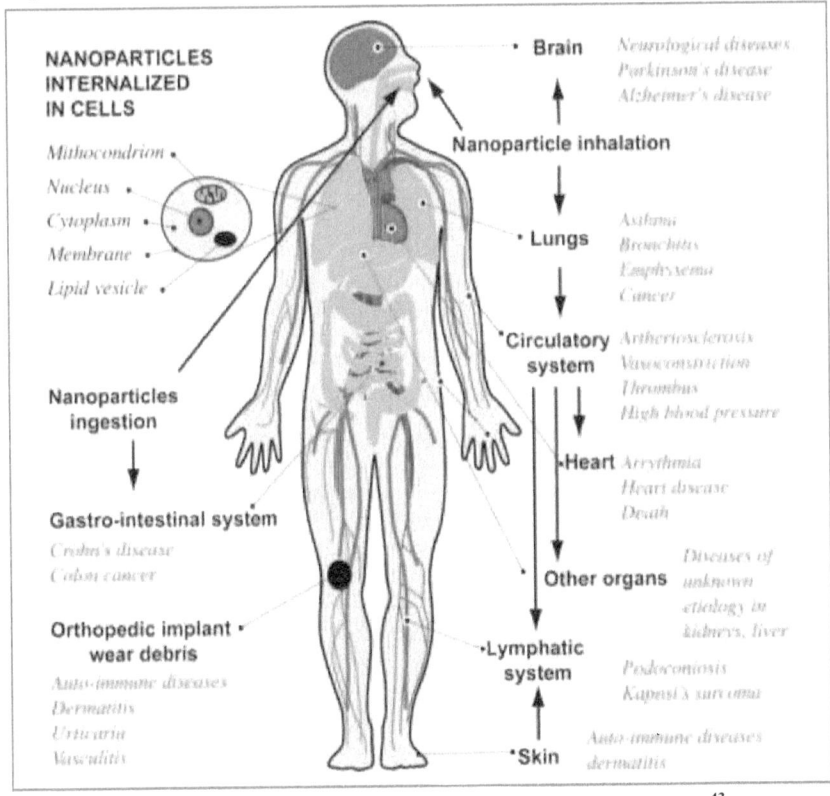

Fig. 5.1 Human exposure to nanoparticles[43]

- ❖ Journal of Molecular Cell Biology published an article disclosing Ployamidoamine dendrimers (PAMAMs) nanoparticles having application in medicine was found to affect lung by programmed cell death known as autophagic cell death.
- ❖ Incidental administrations of insoluble nanoparticles of equivalent mass doses in rats were found to be more potent than larger counterpart causing lung tumors and pulmonary inflammation.
- ❖ Changes in the chemical composition, size of particles, and crystal structure can affect their cytotoxicity and oxidant generation properties as observed in studies relating to animals, cell cultures and cell-free systems.

According to the 2006 survey, two-thirds of the businessmen in nanotechnology believed that unknown risks lie on the exposure of nano particles to the public, workforce and environment and 97% of them believed that the government should take measures to address nanotechnology related environmental and health risks[44]. Very little information is available on the Nanomaterial exposure routes, levels, and related toxicity and even that information are based on the studies of ultrafine particles (<100nm). Those who are working with nanomaterials are at the higher risk of being exposed to engineered nanomaterials of unknown nature.

In short, nanoparticles may be summarized as follows: cytokine production, inflammation, cytoskeletal changes, oxidative stress, altered vesicular trafficking, apoptosis and alteration in gene expression and response to cell signaling to various types of nanoparticles. The targets within cells include the lysosomes. Change in lysosomal permeability and the subsequent release of lysosomal enzymes is one of the mechanisms involved in the induction of alveolar macrophages by silica microparticles (Thibodeau et al., 2004). It was found that multi-walled CNT inhaled by mice suffered from systemic immune suppressed functioning (Mitchell et al., 2007)

Nanomaterials like fullerene C60 water suspension have antibacterial effect; show cytotoxicity to human cell lines; taken up by human keratinocytes; stabilizes proteins, Hydroxylated fullerene results in oxidative eukaryotic cell damage, Metallofullerene accumulates in rat livers, Anatase (TiO_2) shows antibacterial activity and pulmonary inflammation in rodents[40].

5.4 Nanoparticles in Aquatic and Terrestrial Environments

For a while Engineered nanoparticles (ENP) is most often used in every material/product we met on our day to day life and even in some industrial processes. These ENPs will continue to increase in most of the products we encounter and thereby leading to increased discharge of nanoparticles into the water and soil during their application and hence an increasing concern about the environment of aquatic and terrestrial organisms and their impact is raised. Once in the environment, ENPs may be absorbed by plants and microorganism through the passive or active uptake and may cause effects in humans and animals. Because of their large surface areas and highly reactive nature, they can carry toxic materials such as heavy metals and lipophilic pollutants.

Nanoparticles in terrestrial environments

A limited data are available on the toxicity and behaviour of nanoparticles in terrestrial systems. This is because it is difficult to assess the amount of engineered nanoparticles present in the environs of natural nanoparticles in the soil. An in vitro analysis of the culture of terrestrial species of some nanoparticles has shown adverse effects but there is no in vivo evidence proving it.

Nanoparticles are naturally present in terrestrial ecosystems, such as organic matter, iron oxides, clays, and other minerals which are important for biogeochemical processes. Thus, any engineered NPs may interact with natural NPs and hinders the natural development or behavior of soil.

Some NPs may enter the plants through the cell walls of the roots. Cell walls being semi-permeable with a pore size of 5 to 20nm easily permit nanoparticles to pass through them and reach plasma membrane. Evidence disclosing NPs entering into cells through ion channels and embedded transport carrier proteins are available. It was found that due to production of reactive oxygen species (ROS), NPs may hinder with metabolic processes of plants. NPs in the atmosphere can deposit over the leaf surface and through stomata they may enter into the cell. Accordingly, the plants with a larger leaf area and larger stomatal cross section may have the highest penetration rate for atmospheric NPs. Thus, the accumulation of NPs may disturb the gas exchange through stomatal tissues, resulting in the negative effects on the physiology of plants and foliar heating. Aluminium nanoparticles and carbon nanotubes were found to hinder growth of the root in various plant species. Carbon black has been identified to affect the successive fertilization rate of marine seaweed as they accumulate in the sperm cells. Impact of nanoparticles on various food crops have been recently identified among which Carbon NPs deposits over crops results in low rice yields and made wheat susceptible to pollutant attack. Deposition of NPs on leaves may also reduce light penetration and thereby reducing the rate of photosynthesis. Thus nanoparticles require attention as they pose a risk in towards food production.

Also, NPs are most widely used as microbicides for pathogenic bacteria, but they also have an effect on beneficial soil microbes that help in promoting environmental processes, such as plant growth, rotation of element via biogeochemical cycles and removal of pollutants. Certain nanoparticles like that of Ag, ZnO and CuO are toxic to both pathogenic bacteria (e.g., *Staphylococcus aureus* and *Escherichia coli*) and beneficial microbes (e.g., Pseudomonas putida - helps in bioremediation). As they have higher surface area, they show greater toxicity than their equivalent bulk materials. Therefore, terrestrial systems being the largest sink for NPs, a great deal testing techniques and research is required to understand mode of action.

Nano ZnOs were found to be toxic to vertebrates and bacteria. It also hinders germination of seeds and reduces the growth of root in plants. Zinc oxide inhibits the reproducibility and growth of roundworms. Nanosized TiO_2 found to have good photocatalytic properties and hence finds application in industry. Nevertheless, they possess ill effects on bacteria in soil, even when light-generated Reactive oxygen species (ROS) is not present TiO_2-NPs (15–20nm) reduced extractable soil DNA, substrate-induced respiration and triggered soil bacterial community shifts and diversity decline.

Recent findings on terrestrial invertebrates have shown upon exposure to nanoparticles of Zinc and engineered oxidized copper and TiO_2, an increase in metallothione in gene expression is being observed.

Nanoparticles in aquatic environments:

Industrial products and wastes tend to end up in waterways, increasing the possibility of ENPs contamination in water. Over there, ENP encounters aquatic organisms, thereby entering into their body via epithelial cells in gut, gills, cornea, skin etc. or by direct ingestion. The fish's mucus layer also uptakes the nanoparticles to some extent. Carbon nanotubes (CNT), gets readily absorbed in the gill mucus of trout. Various cells take up the same kind of nanoparticles through varied pathways.

Uptake of exposed engineered nanoparticles in aquatic organisms is both via ingestion and water and through gills and gut as they are prime tissues and organs being exposed to ENP. An important criteria for uptake of environmental pollutants, is the bioavailability of nanoparticles. Changes in pH, organic matter and ions present will characterize the bioavailability of NPs. It is also noted that salinity increases aggregation of ENP. Although these aggregates will also be bioavailable to a lesser extent as observed with C60 and TiO_2 in Daphnia. Most chemical compounds get absorbed in cellular tissues, but on the contrary CNT residue was found in the guts of the organism. This indicates that CNT are not readily absorbed into tissues of oligochaetes (Petersen et al., 2008).

Silver nanoparticles of 10-200nm in algae *Chlamydomonas reinhardtii* inhibits photosynthesis, 45-200nm of Ag nanoparticles in Zebrafish (*Danio rerio*) results in alteration of gene expression. ZnO reduces development, hatching and mortality in zebrafish embryo; 30nm of TiO_2 in D. magna shows behavioural and physiological alterations and modification in gills, liver, intestine. CdTe Quantum Dots in Pseudokirchneriella subcapitata inhibits growth[45].

5.5 Nano-particles in Atmosphere

Our everyday environment may be surrounded by nearly 20,000 to 100,000 nanoparticles per cubic centimetre of both natural and engineered nanoparticles in the atmosphere. The particulate matter (PM) are either directly released or they are formed by the reaction product in the atmosphere. It is a heterogeneous solid and/or liquid material (with the exception of pure water) present in a suspension into the atmosphere.

Table 5.2: Sources of atmospheric nanoparticles

Natural Sources	Anthropogenic Sources
Atmospheric formations	Vehicles (petrol, diesel, alternative fuels)
Ocean and water evaporation (Sea spray)	Trains, ships, airplanes
Dust storm	Power plants
Erosion	Incinerators
Forest fires	Various processes (smelting, heating, welding)
Volcanic eruption	
Viruses and Bacteria	Black soot and mineral powders
Photochemical reaction	Functionalized fullerenes
	Building demolition
	Commercial productions

In remote sites, it was found that the number of newly formed particles and the production rate of sulphuric acid were linearly dependent. Though no such data relating to the formation of new particles form the binary nucleation or any third species resulting in ternary nucleation is available. The mechanisms involved in atmospheric nanoparticles formation and growth in the forest is not clear. Roughly about $1-2\times10^3 cm^{-3}$ aerosol particles are emitted into the atmosphere during the spring to autumn from the forest. Near sea, the possible formation of particle may be due to seawater bubble-burst process, wherein ternary nucleation of unknown particles help vapours can condense upon them. Particles of size 10nm are formed in aerosols of sea salt, which are originated by breaking waves. Around 5% to 95% of nuclei particles in marine regions is estimated to form from sea salt flux. The concentration of particles in marine region and forest/rural region is between 10^2 - 10^3 particles/cm^3, and 10^3 -10^4 particles/cm^3, respectively.

Anthropogenic sources form the basis of nanoparticles emissions in urban areas with vehicular exhaust being the strongest contributors of nanoparticles. About 167.7×10^3 particles/cm^3 of nanoparticles are emitted by road tunnels alone. These particles can trap pollutants from vehicular exhaust thereby limiting the exhaust emissions from getting dispersed off. The vehicular emission contributes nearly 86% of total nanoparticle emission in urban regions. Vehicular exhausts that are emitted directly from the engines form the primary particles, mostly containing 30-500nm agglomerates of solid carbon containing material like metallic ash and sulphur compounds and condensed or adsorbed hydrocarbons. When the hot exhaust gases from vehicle cools and condensate (nucleates) form particles of size smaller than 30nm, they are called secondary particles and they mainly comprise of hydrated sulphuric acid and hydrocarbons.

Apart from vehicular exhaust, industrial sources also contribute to atmospheric nanoparticles; though the contribution is much lower (2% of the total particle number). They are mainly from incinerators, power plants, or other industrial processes like welding, smelting or heating operations. In recent years greater advancement is observed in nanoscience as it's used in increased in various fields such as medicine, electronics, material sciences, energy storage, etc. Thus, engineered nanoparticles other than vehicular exhaust and industrial emission also form the major source of atmospheric emission[46].

Engineered (or manufactured) nanoparticles

Engineered nanoparticles are tailor made nanoparticles, manufactured to meet certain specific properties or composition. They are different from air pollution, natural nanoparticles. The British Standards Institution (BSI, 2007), have classified engineered nanomaterials into four hazardous groups:

1. **Fibrous nanomaterials**; They are insoluble particles with >3:1 length to diameter ratio.
   a. Carbon nanotubes (CNTs): very commonly used as it is 100 times stronger than steel and comparatively, weighs very less.
   b. Nanowires: they comprise of carbon, sulphides, oxides, metals and possess complete or partial electric conductivity
2. **CMAR Nanomaterials (carcinogenic, mutagenic, asthmagenic or reproductive toxins):** Nanopartilces made up of nickel, arsenic and cadmium.
3. **Insoluble nanomaterials**:
   a. Quantum dots (QDs): They are semi-conductors with unique optical properties, mostly used in diagnostic medical imaging.
   b. Fullerenes: C_{60} is commonly used fullerenes as its shape do not deform even after application of extreme pressures. They have clatherate like structure helpful in drug delivery.
   c. Titanium Dioxide (TiO_2): Nanosized TiO_2 finds extensive application in sunscreen and paints as they are transparent to visible light.
4. **Soluble nanomaterials:** Most engineered nanomaterials have poor solubility[47].

Fate of Engineered Nano-particles (ENP) in atmosphere

The factors responsible for the fate of ENPs in the air are: the length of time particles remain suspended, their chemistry with other suspended particles in the atmosphere and the distance they move in the air. The processes that are necessary to understand the behaviour of ENPs in the atmosphere are diffused, agglomeration, wet and dry deposition and settling down by gravity. Their dynamics are similar to the ultrafine particles, though there are few differences like the ultrafine particle are coated and hence they do not agglomerate like nanoparticles. Nanoparticles that agglomerate rapidly have reduced residence time as they agglomerate and settle down. While for non-coated nanoparticles their residence time is unpredictable as they do not agglomerate and settle down fast it is believed that the nanoparticles once deposited will probably not get suspended or aerosolized back into the atmosphere.

Most of the ENPs are photoactive, but susceptibility to photodegradation in the atmosphere is unknown. Still, the chemistry of interaction of ENPs with other elements present in the atmosphere is to unfurl[48].

Nanoparticles are naturally formed in the air via atmospheric formation, volcanic eruptions, vegetation sprays, forest fires and sea sprays. The mechanism of atmospheric formations was ideally considered as homogenous nucleation involving photochemically induced nucleation, and/or nucleation through gas-to particle conversion of semi-volatile organic aerosols. Other means of formation involves a few different nucleation mechanisms: (i) binary nucleation involving the co-condensation of sulphuric acid and water; or (ii) ternary nucleation involving a third substance, like an organic acid or ammonia which can enhance the nucleation rates thus said binary system.

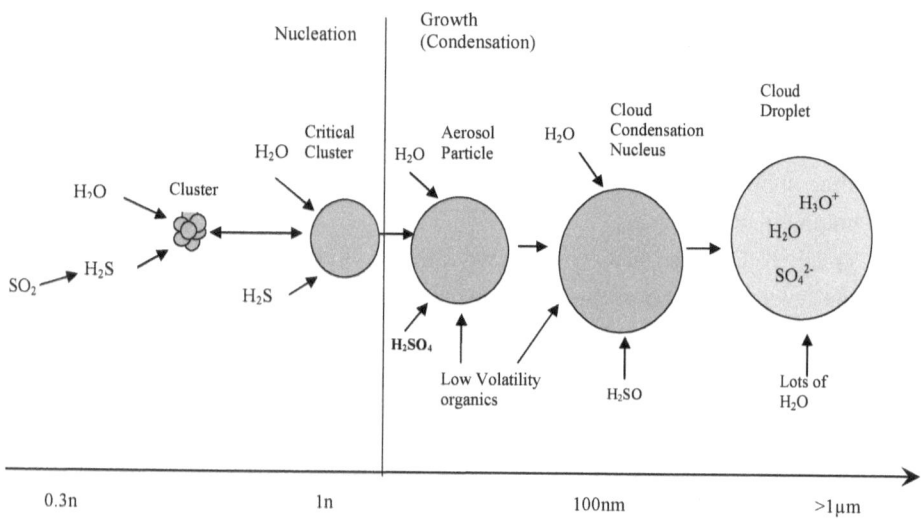

Fig.5.2: A schematic representation of nucleation and subsequent growth process of atmospheric binary homogeneous nucleation of H_2SO_4 and H_2O[49]

The finest nanoparticles of size less than 20nm, will scarcely have a primary emission source. They often originate from organic (VOC) and inorganic (SO_2, NH_3) gaseous precursors into the atmosphere, by photochemical processes or by cooling down of substances emitted from combustion sources. The mid-sized nanoparticles otherwise called as Aitken particles range of 20-100nm. They are generated by the agglomeration of fine suspended particles or when condensation takes place on nucleation particles, and also by direct emission of carbon containing particles from diesel engines (soot particles). The Aitken particles are mainly made of organic sulfate and nitrate species. The particles in accumulation mode of size 100-1000nm, are often made of combustion particles and secondary inorganic particles. Thus, they are basically generated by a secondary process via condensation processes and coagulation of finer particles, though primary emission also shows a contribution. The particle in coarser mode of size 1-10μm are mainly primary particles formed by various natural and anthropogenic processes, while the secondary particles are formed by either condensation or reaction of saturated organic and inorganic gases with pre-existing condensation nuclei. Trace metals having mass concentration are ubiquitous in urban environments. For example elements like a V, Ni, Co, Sb, Cr, Fe, Mn, Cu, Zn, As and Sn are emitted by fossil combustion and metallurgical activity in the atmosphere while the traffic emission involves trace element like Ba, Pb, Cu, Cr, Sn, Sb and Zr. Atmospheric metalliferous particles which are present in nanometric size reacts easily with human fluids and get transported over distances of hundreds of kilometers smoothly.

5.6 Health Impacts of Nanoparticles

After being exposed to an aqueous and aerial environment nanosized particles enter the human body via lungs, skin and gastrointestinal tract. Because of their smaller size, unlike coarse particles, they behave differently. When inhaled their small size, permit them to get deposit over the lungs and penetrate through alveolar epithelium. From there they infiltrate into the pulmonary interstitial and vascular space and finally to the blood stream. The nanoparticles through the systemic circuit may enter into organs like the liver and may translocate to the central nervous system, the brain. They carry a large surface area and hence are toxic and more reactive, leading to adverse health effects such as stress, oxidation, pulmonary inflammation and cardiovascular disorder.

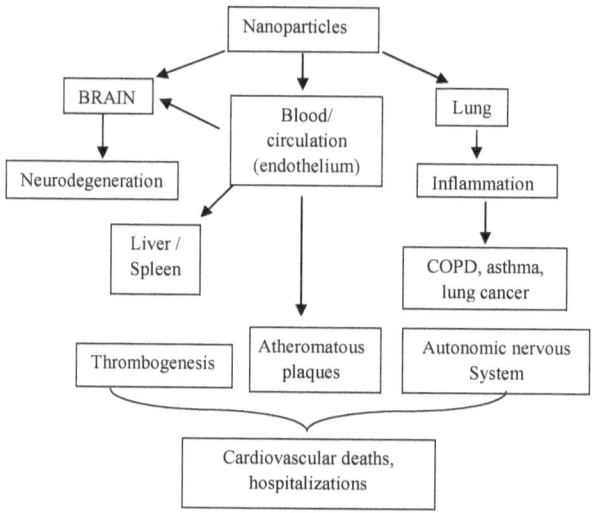

Fig. 5.3 Systemic health effects

Source: Systemic health effects of atmospheric nanoparticles[48] (Terzano et al. (2010)

Environmental impacts include reduced visibility and climatic changes.

5.7 Toxicological Properties of Nanoparticles and Nanotubes

Nanotechnology used in almost every field and humans are daily exposed to airborne nanomaterials that are found in drugs, smoke, paints, coatings, sunscreens, cosmetics, shampoos, soaps, detergents, tennis rackets, catalysts, video screens, and concrete. The nanoparticles being small in size easily cross through the biological membranes and enter into the blood streams, cells, tissues and organs through inhalation or ingestion and skin penetration. Injuries like severe sunburn, shaving wounds or eczema may increase the rate of nanoparticles uptake by skin as the broken skin act as an inlet to these particles. They enter into the blood stream via inhalation and ingestion and from there on they can be transported throughout the body and absorbed by organs, tissue and cells, resulting in increased oxidative stress, inflammatory cytokine production and cell death. Nanoparticle, unlike larger particles, can enter cell mitochondria and have the potential cause of mutation in DNA and stimulate major structural damage to mitochondria and can even cause cell death. Therefore, size is a major criterion in deducing the toxicity of a particle. Other variables should also take into consideration while working with nanomaterials and these include shape, material, surface, coating, charge, dispersion, aggregation, agglomeration, matrix, and concentration to determine its exact toxicological effect, which makes it more complex. The table below summarizes the observed biological effect vis-à-vis the physicochemical properties and the types of nanomaterials.

Physicochemical properties of Nanomaterials show certain biological effects as follows:[50]

Size: 15-50nm gold has the potential of pass through the blood-brain barrier while 15nm gold nanoparticles has the ability to pass through blood, liver, lung, spleen, kidney, brain, heart, stomach in mice causing bio-distribution of nanoparticles.

Shape: Spherical gold Nps induces higher uptake by Hela cells while rod - shaped gold Nps showed less uptake. Filamentous micelle of nanoparticles shows more efficient drug delivery than their counterparts in rat and mice.

Surface area / Volume ratio: TiO_2 of $300 cm^2$ surface area induces increased lymph-node burdens and inflammation.

Chemical Composition: Incorporation of 1% (w/w) manganese doping into Titania particles shows increased UVA absorption and reduction in free radical generation via surface reactions. Quantum dot core metalloid complexes of selenium, shows a marked impact on the local ecosystem resulted from elevated environmental concentrations of Se.

Surface Charge: Neutral NPs and low concentration anionic NP possess Drug delivery applications to brain in rats.

Routes of exposure
Inhalation of nanoparticles:

Through inhalation the nanoparticles enter the respiratory system, depositing within nasopharyngeal, tracheobronchial and alveolar regions of the lung depending on the size variation. Once in the lungs, these nanoparticles start interacting with different biological systems causing detrimental effects. Like, they can enter bone marrow, lymph nodes, spleen and heart through blood and lymph. Studies have proven the association of inhaled nanoparticles with cardiovascular disorder like cardiac rhythm disturbances and coagulation. Mixed carbon nanoparticles and nanotubes MWCNT and SWCNT were found to stimulate in vitro aggregation of platelet and in rat carotid artery, they were found to accelerate the rate of vascular thrombosis. Nanoparticles can also directly induce cytotoxic morphological changes in induction of proinflammatory responses, human umbilical vein endothelial cells, inhibition of cell growth and reduction of endothelial nitric oxide synthase.

Inhalation and instillation experiments of nanoparticles ingestion were found to affect the sensory nerve endings found in the epithelia airway, followed by ganglia and the central nervous system through the axons. Alveolar macrophages absorb ultrafine silver particles and they persist, there for nearly seven days and showed cytotoxicity. C_{60} fullerenes stimulate oxidative stress in the brain via the olfactory bulb of largemouth bass.

The airline crews and hardware engineering may show concern towards nanoparticles released in the confined environment of aerospace and computing technology. In addition, nanoparticles are used as drug carrier systems in aerosol therapy wherein liquid medication turns into mist easy for inhalation have shown to stimulate increased lung toxicity compared to conventional particles. In experimental animals it was found that nanoparticles with size variation induce inflammatory reactions in the lungs. In fact, an increased oxidative stress was observed with increased surface area of nanoparticles.

Skin penetration of nanoparticles:

The skin is the largest organ of the body and functions as the first-line of defense between the external environment and the internal organs of the human body. The skin contains three layers: the epidermis, the dermis and the subcutaneous layer. Interaction of nanoparticles with skin can be a result of intentional topical application creams and other products containing nano silver, quantum dots, nanotitania and nano zinc oxide. Intercellular, intracellular, and follicular penetration are the three ways through which skin uptake the matter. Diffusion dominates in uptake process. Lipophillic materials having a low molecular weight, diffuse through the lipid-rich intercellular space of the stratum corneum. Materials that penetrate the stratum corneum into the stratum granulosum can induce the resident keratinocytes to release pro-inflammatory cytokines. Materials that penetrate to the stratum spinosum, which contains Langerhans cells (dendritic cells of the immune system), can initiate an immunological response. This is mediated by the Langerhans cells, which can become antigen-presenting cells and can interact with T-cells. Once these materials enter the stratum granulosum or stratum spinosum part of the skin, they can very easily get an access through the circulatory and lymphatic systems. Dry powder of engineered nanoparticles enters into the body though inhalation exposure, whereas those in liquids, liquid dispersed ENMs pose a greater risk for dermal exposure. Single Walled CNTs shows oxidative stress in the skin and skin thickening.

Prolonged dermal exposure of sunscreen containing titanium nanoparticles was found to penetrate into the epidermis and dermis. In vitro experiments of hair follicles of massaged porcine skin showed deeper penetration of nanoparticles with a dye and remained for a long time in human skin in vivo than the dye in solution. Ultraviolet radiation enhanced penetration which may cause a concern with nanoscale sunscreens. Cadmium from cadmium-containing quantum dots was seen in the liver, spleen, and heart; However, ICP-MS method through which it was determined did not give a clear picture on the mode of penetration of cadmium. Such toxicity of surface treatments can be minimized by trapping the free radicals of reactive oxygen species (ROS).

Ingestion of nanoparticles

Nanoparticles can be ingested directly from food such as titanium oxide (as colourant), water, pharmaceuticals, or cosmetics like toothpaste, dental prosthesis debris. In developed countries, it is estimated about 10^{12} of nanoparticles is consumed by a person in a single day through food, which mainly comprises Titania and mixed Silicates nanoparticles. They get accumulated in macrophages and posses very slow degradation rate. A little part of inhaled nanoparticles was also found to enter the gastrointestinal tract. Nanoparticles' surface chemistry and charge, size, length of administration, and dose determine the extent of absorption in the gastrointestinal tract. Most of the nanoparticles get eliminated from the intestinal tract, though some get translocated to spleen, blood, liver, lymph nodes, bone marrow, kidneys, brain, and lungs, and can also be found in the stomach and small intestine. Its ways of translocation from the GI tract to other organs and blood are not identified, however, some studies have proven intestinal absorption of particles is followed by liver clearance before they reach the general circulation and the kidneys.

5.8 Nanotube Toxicology

Carbon nanotubes are made up of graphene sheets in a rolled up manner. They are classified as single walled (SWCNTs) which comprises a single layer of rolled up graphene sheet of about 0.4nm diameter and micrometers length or multiwalled carbon nanotubes (MWCNTs) which has multiple layer of graphene sheets all rolled up in a concentric cylindrical tube with diameter upto 100nm. Due to their unique structure, they are used in energy storage, boat hulls, automotive parts, sporting goods, thin-film electronics, water filters, actuators coatings, electromagnetic shields, etc. CNTs are not extensively used in biological or medical application because of its inert and hydrophobic nature, though they are used in various biomedical devices and therapies. One such example is its application in drug delivery to cancer cells owing to strong optical absorption in the near IR region and high surface area. But very less data are available on its toxic effect. However, CNTs were found to be toxic to lung and embryos. This biological toxicity can be reduced by covalent modification and adding surfactants to the CNTs.

CNT's may very quickly be removed from the blood and get confined in the liver, spleen and lung. On intravenous or intraperitoneal administration, CNT most dominantly accumulates in the liver. ss-DNA-MWCNTs were found to induce inflammation and oxidative stress in plasma and liver. At higher concentration pristine CNT exposure has shown to cause less cytotoxicity while the functionalized CNT for drug delivery has not shown any toxicity so far[51].

5.9 Development of Safe Nanotechnology

Nanotechnology is used to create many new materials and devices possessing diverse applications, such as in energy production and medicine, electronics. Its astounding fact is that just inducing minor changes in its size range, nanoparticles exhibit completely new characteristics. For e.g., TiO_2 is white in color but on reducing its size to <50nm, it becomes colorless. As we know everything comes with a cost, and so its potential health and environmental issues should be addressed before developing a new product. A variety of physiochemical characteristics are known to have a toxic effect and transport nanomaterials in the body and the environment. So Nanomaterials – a boon or bane? This question still prevails. As much uncertainty exists in regards to the impact of nanotechnology this technology in the future has the potential; to be inclined towards bane. But this can be checked by developing safe nanotechnology and several organizations are opting for safe technologies by enhancing safety for workers, consumers, waste handlers, and for protecting the environment.

Inculcating safe nanotechnology has been an integral part of any novel nanotechnology development. The crucial challenges that the today's world is facing in regards to safe nanotechnology is that inappropriate or laborious tools are available for the safety assessment of engineered nanomaterials. Current or known resources or test methods are not applicable to assess the safety of the numerous new nanomaterials that are emerging very rapidly. This calls for new safety assessment standards and methods for safe development in the coming years.

Also, there is very less information on fire prediction and explosion risk associated nanomaterial powders. Because of its larger surface area and smaller size, nanosized combustible material exhibits a higher risk than its coarser counterpart with a similar mass concentration. The composition and structure of some nanomaterials may start to catalyse a reaction which otherwise would not have been expected based on their chemical composition.

Certain rigid products based on nanotechnology, such as nano composites, materials comprised of nano-structures, such as integrated circuits, possess zero risk of exposure during their use and handling as these materials are of non-inhalable size. However, cutting or grinding these products and while manufacturing, there is a complete possibility of releasing respirable-size (<PM10) nanomaterials that can be inhaled causing toxicity. Also, maintenance of production systems (including cleaning and disposal of materials) may also lead to exposure to nanoparticles if deposited nanomaterials are disturbed. The United States National Institute for Occupational Safety and Health have reported that the following workplace activities can increase the risk of exposure to nanoparticles[53]

- Working with nanomaterials in liquid media without adequate protection (e.g., gloves)
- Working with nanomaterials in liquid during pouring or mixing operations, or where a high degree of agitation is involved
- Producing nanoparticles in open systems
- Handling (e.g., weighing, blending, spraying) nanomaterials powders
- Maintenance on equipment and processes used to produce or fabricate nanomaterials and the cleaning-up of spills and waste material containing nanomaterials
- Cleaning of dust collection systems used to capture nanoparticles
- Machining, sanding, drilling, or other mechanical disruptions of materials containing nanoparticles

There are five main challenges that have to overcome to focus the research towards safe nanotechnology. They are

1. **To develop instruments that can determine the nano-materials exposure in air and water:** Due to the diverse nature of nanotechnologies, exposure of nano-materials in the environment will vary widely. It is therefore important to develop instruments that will help in detecting any type of nanomaterials present in the atmosphere and water also can estimate its potential health and environmental impact.

2. **To develop methods that can help in evaluating nano-materials' toxicity.** It requires validated screening tests, finding and developing a feasible substitute to in vivo tests, and assessing the fibre-shaped nano-particles' toxicity.
3. **To develop robust ways of determining the effects of nanomaterials throughout its life cycle, i.e.,** from the manufacturing stage to its disposal.
4. **To develop planned programmes that will focus on probable risks**[53]

Precautionary Measures that can be taken to ensure safety[52]:

- As it is known that only limited knowledge is available on health risks associated with nanomaterials, it is sensible to take measures to reduce worker exposure
- The exposure to airborne nano-aerosols can be controlled using techniques analogous to those used in minimizing exposure to non-nano aerosols.
- The Risk management program should be implemented in places where nanomaterials exposure exists. Elements that can be included in such a program are:
  - ✓ Evaluating the hazard generated due to exposure nanomaterial based on the data available (physical and chemical property, toxicology or health-effects).
  - ✓ Determining the worker's job task to assess the potential for exposure.
  - ✓ Generating awareness among workers regarding the proper handling of nanomaterials (e.g., good work practices)
  - ✓ Proper criteria and procedures should be established for installing and evaluating control equipments (e.g., exhaust ventilation) at places where nanomaterials exposure might occur.
  - ✓ To develop procedures for evaluating the need for and selecting proper personal protective equipment (e.g., clothing, gloves, respirators).
  - ✓ Systematically evaluating exposures to ensure that control measures are working properly and that workers are being provided the appropriate personal protective equipment.
- Types of clothes/apparel that can be draped to prevent dermal exposure are not known. However, some clothing standards incorporate testing with nanometer-sized particles and therefore provide some indication of the effectiveness of protective clothing.
- Respirators may be used when no adequate control techniques are available to prevent exposures. Exposure limit applicable for large particles of similar composition is followed for nanoparticles too as there is no specific limit available for exposures to engineered nanoparticles. Though it will not ensure

many of the safety concerns. The decision to use respiratory protection should be based on professional judgment that takes into account toxicity information, exposure measurement data and the frequency and likelihood of the worker's exposure.

References:

1. C. Scherer and A. M. Figueiredo Neto, "Ferrofluids: Properties and Applications," Brazilian J.Phys., Vol. 35(3A), pp. 718-727, 2005.
2. Mahmoudi, M.; Simchi, A.; Imani, M.; Stroeve, P.; Sohrabi, A., "Templated growth of superparamagnetic iron oxide nanoparticles by temperature programming in the presence of poly(vinyl alcohol)", Thin Solid Films, vol. 518(15), pp. 4281-4289, 2010.
3. Deepa Thapa, V.R Palkar, M.B Kurup, S.K Malik, "Properties of magnetite nanoparticles synthesized through a novel chemical route," Materials Letters Volume 58, Issue 21, , Pages 2692–2694. August 2004
4. L. Vékás, "Magnetic nanofluids properties and some applications," Rom. Journ. Phys., vol. 49, pp. 707-721, 2004.
5. Shouhu Xuan, Lingyun Hao, Wanquan Jiang, Xinglong Gong, Yuan Hu, Zuyao Chen, "Preparation of water-soluble magnetite nanocrystals through hydrothermal approach," Journal of Magnetism and Magnetic Materials, vol. 308, pp. 210-213, 2006.
6. Ye XR, Daraio C, Wang C, Talbot JB, Jin S., "Room temperature solvent-free synthesis of monodisperse magnetite nanocrystals." Journal of Nanoscience and Nanotechnology, vol. 6(3), pp. 852-856, 2006.
7. Ana L. Daniel-da-Silva, T. Trindade, Brian J. Goodfellow, Benilde F. O. Costa, Rui N. Correia, and Ana M. Gil, "In Situ Synthesis of Magnetite Nanoparticles in Carrageenan Gels," Biomacromolecules, vol. 8, pp. 22350-2357, 2007.
8. Ki Do Kim, Sung Soo Kim, Yong-Ho Choa, and Hee Taik Kim, "Formation and Surface Modification of Fe_3O_4 Nanoparticles by Co-precipitation and Sol-gel Method," J. Ind. Eng. Chem, vol. 13(7), pp. 1137-1141, 2007.
9. N. Mizutani, T. Iwasaki, S. Watano, T. Yanagida, H. Tanaka, T. Kawai, "Effect of ferrous/ferric ions molar ratio on reaction mechanism for hydrothermal synthesis of magnetite nanoparticles," Bull. Mater. Sci, vol. 31(5), pp. 713-717, 2008.
10. Marie L. Garcia, Olin H. Bray, Fundamentals of Technology Roadmapping, Sandia National Laboratories, SAND97-0665 April 1997.
11. C. T. Lee, Ci-Ling Pan, Photovoltaic technology with self-assembled nanostructures of silicon quantum-dots in mesoporous silica, NSC Nanoscience Program, Aug. 1, 2008- July 31, 2011
12. Nanomanufacturing for Energy Efficiency Workshop report, Industrial; technology program, U. S. Department of Energy 2007.

13. Prash Makaram, Nanomaterials: Nanotechnology on Earth and Beyond Series, http://futurehumanevolution.com/nanotechnology-on-earth-and-beyond-nanomaterials.
14. M.H. Fulekar, Nanotechnology: Importance and Applications, I. K. International Pvt Ltd, 2010.
15. Review of international nanotechnology developments and policy concerns, Project Report No. 2006ST21:D1, The Energy and Resource Institute October 2009.
16. Thomas Heinze, Stefan Kuhlmann, Across institutional boundaries? Research collaboration in German public sector nanoscience, Research Policy 37 (2008) 888–899, April 2008.
17. Thomas Heinze, Emergence of Nano S&T in Germany. Network Formation and Company Performance, Fraunhofer ISI Discussion Papers Innovation System and Policy Analysis, ISSN 1612-1430, Karlsruhe, April 2006.
18. Chul-joo Lee, Dietram A. Scheufele, The influence of knowledge and deference toward scientific authority: a media effects model for public attitudes toward nanotechnology, J&MC Quarterly Vol. 83, No. 4, Winter 819-83 2006.
19. David M Berube, Brenton Faber and Dietram A. Scheufele Communicating Risk in the 21st Century – A case of nanotechnology, Risk Communications, Feb 2010.
20. Lars Guenther & Georg Ruhrmann, Science Journalists' selection criteria and depiction of nanotechnology in German media, Journal of Science Communication JCOM 12(03), Autumn/Winter 2013.
21. Department of Science & Nanotechnology: Nano Mission-Objectives (http://nanomission.gov.in/).
22. Nico Jaspers, International Nanotechnology Policy and Regulation - Case Study: INDIA, June 2010.
23. European Commission: Research & Innovation: Key areas of the European Strategy and the Action Plan. (http://ec.europa.eu/research/industrial_technologies/policy-key-areas_en.html)
24. National Nanotechnology Initiative (http://www.nano.gov).
25. John F. Sargent Jr., Nanotechnology: A Policy Primer, CRS Report for Congress, Dec 2013.
26. Jarvis, S.L., Richmond, N., "Regulation and Governance of Nanotechnology in China: Regulatory Challenges and Effectiveness" European Journal of Law and Technology, Vol. 2, No.3, 2011.

27. Georgia Miller, Nanotechnology, Ubiquitous Computing and the Internet of Things: Challenges to rights to privacy and data protection draft report to the council of Europe, September 2013.
28. Robert McGinn, Ethical Responsibilities of Nanotechnology Researchers: A Short Guide, Nanoethics 4:1–12 (2010).
29. Nanotechnology: A Realistic Market Assessment - NAN031F, November 2014
30. The Great Eight - Trillion-Dollar Growth Trends to 2020 - Bain & Company, Inc, 2011.
31. Quantum dot (QD) market global analysis, growth, trends, opportunities, size, share and forecast through 2020, Allied market research, April 2014.
32. Expanding Applications to Drive Growth in the Global Nanotechnology Market, According to New Report by Global Industry Analysts, Inc., November 12, 2012.
33. Roya Naseri, Reza Davoodi, Commercialization of Nanotechnology in Developing Countries, 3rd International Conference on Information and Financial Engineering, IPEDR, vol.12, 2011.
34. Strategic Priorities for Nanotechnology Program, Kingdom of Saudi Arabia, King Abdulaziz City for Science and Technology, Ministry of Economy and Planning.
35. WIPO - What is Intellectual Property?
http://www.wipo.int/edocs/pubdocs/en/intproperty/450/wipo_pub_450.pdf
36. Terry K. Tullis, Current Intellectual Property Issues in Nanotechnology, UCLA Journal of Law & Technology, 2004.
37. The Patent Act, 1970, Intellectual Property, India.
38. Protecting your Inventions Abroad: Frequently Asked Questions About the Patent Cooperation Treaty (PCT), April 2015
(http://www.wipo.int/pct/en/faqs/faqs.html).
39. B. Zhang, H.Misak, P.S. Dhanasekaran, D. Kalla, R. Asmatulu, Environmental Impacts of Nanotechnology and Its Products, Proceedings of the 2011 Midwest Section Conference of the American Society for Engineering Education.
40. Nanotechnologies in the 21st century: Nanomaterials - health and environmental concerns: Issue 2, Environmental European Bureau, July 2009.
41. The disruptive social impacts of nanotechnology, Sep 2006
(http://emergingtech.foe.org.au/152/).
42. Center for Responsible Nanotechnology- Dangers of Molecular Manufacturing (www.crnano.org/dangers.htm).

43. Haseeb Ahmad Khan, Ibrahim Abdulwahid Arif, Toxic Effects of Nanomaterials, Bentham Science Publishers, Jan-2012.
44. John F. Sargent, Nanotechnology and Environmental Health and Safety: Issues for Consideration, Congressional Research Service, Federal Publications, Jan 2011.
45. Karl Fent, Ecotoxicology of Engineered Nanoparticles, F.H. Frimmel, R. Niessner (Eds.), Nanoparticles in the Water Cycle, DOI 10.1007/978-3-642-10318-6_11, 2010.
46. Klara Slezakova, Simone Morais, Maria do Carmo Pereira, Atmospheric Nanoparticles and Their Impacts on Public Health, Chapter 23, DOI: 10.5772/54775, May 2013.
47. Xavier Querol, Andrés Alastuey, Jorge Pey, Mar Viana, Teresa Moreno, MaríaCruz Minguillón, Fulvio Amato, Marco Pandolfi, Noemí Pérez, Michael Cusack, Cristina Reche, Manuel Dall'Osto, Anna Ripoll, Angeliki Karanasiou, Nanoparticles in the atmosphere, Institute of Environmental Assessment and Water Research.
48. Riipinen, I., Yli-Juuti, T., Pierce, J. R., Petaja, T., Worsnop, D. R., Kulmala, M., Donahue, N. M.. The contribution of organics to atmospheric nanoparticle growth. Nature Geoscience: 5: 453–458 2012.
49. J. Curtius, Nucleation of atmospheric particles, European Physics Journal, Conferences 1, 199–209 (2009)
50. Sumit Arora, Jyutika M Rajwade, Kishore M Paknikar, Nanotoxicology and in vitro studies The need of the hour, Toxicology and Applied Pharmacology 2;258(2): 151-65 Jan 2012.
51. Stefano Bellucci, Carbon Nanotubes Toxicity, Ed.), Nanoparticles and Nanodevices in Biological Applications. Lecture Notes in Nanoscale Science and Technology 7, 2009.
52. Approaches to Safe Nanotechnology - Managing the Health and Safety Concerns Associated with Engineered Nanomaterials - Department of Health and Human Services (DHHS) National Institute for Occupational Safety and Health (NIOSH) Publication 2009.
53. Andrew D. Maynard1, Robert J. Aitken, Tilman Butz, Vicki Colvin, Ken Donaldson, Günter Oberdörster, Martin A. Philbert, John Ryan, Anthony Seaton, Vicki Stone, Sally S. Tinkle, Lang Tran, Nigel J. Walker & David B. Warheit, Safe handling of nanotechnology, Nature 444, 267-269, Nov 2006.

I want morebooks!

Buy your books fast and straightforward online - at one of the world's fastest growing online book stores! Environmentally sound due to Print-on-Demand technologies.

Buy your books online at
www.get-morebooks.com

Kaufen Sie Ihre Bücher schnell und unkompliziert online – auf einer der am schnellsten wachsenden Buchhandelsplattformen weltweit!
Dank Print-On-Demand umwelt- und ressourcenschonend produziert.

Bücher schneller online kaufen
www.morebooks.de

OmniScriptum Marketing DEU GmbH
Heinrich-Böcking-Str. 6-8
D - 66121 Saarbrücken
Telefax: +49 681 93 81 567-9

info@omniscriptum.com
www.omniscriptum.com

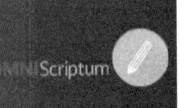

www.ingramcontent.com/pod-product-compliance
Lightning Source LLC
Chambersburg PA
CBHW031535210526
45464CB00003B/1022